PISTOL CREEK, IDAHO
HISTORY AND GEOLOGY
PISTOL CREEK RANCH EVENTS

ANTON P. SOHN MD

TotalRecall Publications, Inc.
1103 Middlecreek
Friendswood, Texas 77546
281-992-3131 281
www.totalrecallpress.com

All rights reserved. Except as permitted under the United States Copyright Act of 1976, No part of this publication may be reproduced, stored in a retrieval system, or transmitted in any form or by any means electronic or mechanical or by photocopying, recording, or otherwise without prior permission of the publisher. Exclusive worldwide content publication/distribution by TotalRecall Publications, Inc.

Copyright © 2023 by Anton P. Sohn MD
Copy Editor: Kristin Sohn Fermoile MD
Graphics Copyright © by Anton P. Sohn MD
Cover: Watercolor of Ranch Runway, Lodge, Barn, and "Abe" by Anton P. Sohn MD
Page 80: Watercolor of Elk Skull on Lodge Deck by Anton P. Sohn MD

ISBN: 978-1-64883-188-1
UPC: 6-43977-21881-0

Library of Congress Control Number: 2023945464

1 2 3 4 5 6 7 8 9 10

FIRST EDITION

Colophon is trademarked

The scanning, uploading and distribution of this book via the Internet or via any other means without the permission of the publisher is illegal and punishable by law. Please purchase only authorized electronic editions, and do not participate in or encourage electronic piracy of copyrighted materials. Your support of the author's rights is appreciated.

THIS BOOK IS DEDICATED TO
Chris and Dave Dewey
Barbara and Pat Hawn
Sonja and Bill Widgren

CONTENTS

ACKNOWLEDGEMENTS AND CONTRIBUTORS VI
ILLUSTRATIONS VIII
INTRODUCTION XI

1. ONE THOUSAND YEARS AT PISTOL CREEK 1
2. PISTOL CREEK RAPIDS CAVE 2
3. SAM HOPKINS AT PISTOL CREEK, 1910 4
4. ELECK WATSON AT PISTOL CREEK, 1914 4
5. BILL MITCHEL/CHARLES MEYERS CABIN, 1920 4
6. FRANK CHURCH "RIVER OF NO RETURN" WILDERNESS CREATED ... 6
7. GEORGE RISLEY'S 144.33 ACRE HOMESTEAD, 1920 7
8. GEORGE RISLEY'S WIDOW SELLS RANCH TO BILL WAYNE, 1950s 8
9. GARDEN CREEK GUARD STATION (GCGS) 8
10. GCGS AND ED HUNTINGTON'S CABIN AT FROG POND, 1938 9
11. PARACHUTE BAILOUT AT PISTOL CREEK, 1943 10
12. ROSS GEILING, 1987, and HIS CABIN at 45 CREEK 12
13. RANCH'S ORIGINAL CABIN & LODGE, 1960 13
14. ADELAIDE AND BILL WAYNE, 1950s 14
15. SULPHUR CREEK TO PISTOL CREEK RANCH FLIGHT, 1950s 14
16. HORNBACKS DEVELOP PISTOL CREEK RANCH, 1957 14
17. FIRST PISTOL CREEK RANCH CABINS, 1960 14
18. BRUCE McKEIGHAN BUYS PISTOL CREEK RANCH AND INCORPORATES MIDDLEFORK RANCH, 1966 14
19. PISTOL CREEK RANCH, 144.33 ACRES, 1914 15
20. ORIGINAL PISTOL CREEK RANCH CABIN OWNERS, 1960s 20
21. PISTOL CREEK RANCH CEMETERY MARKERS 24
22. PISTOL CREEK RANCH CEMETERY RECORDS, 2013 26
23. JOHN LANCASTER PLANE CRASH, JULY 25, 2000 27
24. "BLACK ASH SUNDAY" AUGUST 27, 2000 29
25. POST-FIRE CABINS/LOT OWNERS, 2000 29
26. REBUILDING CABINS, 2001 30
27. CABINS CONSTRUCTED WITH SIPs, 2001 30
28. SIPs TO REBUILD CABINS 31
29. REBUILDING THE RANCH 32
30. POST-FIRE CABINS 33
31. PISTOL CREEK RANCH MANAGERS 36
32. SWAT TEAM AT PISTOL CREEK, 2013 37
33. FISHING 38

34.	HUNTING	39
35.	CAMPING	40
36.	"BALLADS" AND "BLUES"	41
37.	HORSEBACK RIDING	44
38.	GUN RANGE	45
39.	WATER SPORTS AND HIKING	45
40.	SOCIAL ACTIVITIES	46
41.	BIRDS AT THE RANCH	47
42.	HANDCRAFTS	48
43.	CRAFTS	50
44.	WILDLIFE	50
45.	AREA WILDFLOWERS	51
46.	PISTOL CREEK RANCH CABIN OWNERS, 2023	53
47.	MIDDLE FORK GEOLOGY	53
48.	OREGON METEORITE STRIKE	53
49.	SNAKE RIVER PLAIN	53
50.	LAKE MISSOULA	54
51.	IDAHO BATHOLITH	54
52.	IDAHO BATHOLITH MATERIALS	54
53.	IGNEOUS ROCK	54
54.	SALT AND PEPPER ROCKS	54
55.	METAMORPHIC ROCK	55
56.	GEOLOGY DEFINITIONS	55
57.	GETTING ORIENTED AT PISTOL CREEK RANCH	55
58.	TERRACE FORMATION	56
59.	COW CREEK TERRACES	57
60.	PISTOL CREEK RANCH TERRACE	57
61.	MINERALS AT PISTOL CREEK RANCH	59
62.	ROCKS AT PISTOL CREEK RANCH	60
63.	HOT AND COLD SPRINGS NEAR PISTOL CREEK RANCH	62
64.	HOT SPRINGS	63
65.	COLD SPRINGS	63
66.	REMINISCENCES OF PISTOL CREEK RANCH	64
	BIBLIOGRAPHY	76
	INDEX	77

ACKNOWLEDGEMENTS AND CONTRIBUTORS

Note: the following individuals have supplied information for this edition. A book of this scope with detailed history and data on Pistol Creek would not have been possible without the help of the following individuals. Others have supplied important information.

Kristin Sohn Fermoile, my loving daughter, edited this book and reworked my bumbling sentences.

Bruce Moran of TotalRecall Press, who published my earlier books, provided motivation to publish another book.

Robert Boyd (C14)[1] related to us the story of burying Clay Woods' ashes in the cemetery and more.

Rusty and Billy Brace (C10) supplied photographs, information on cabin owners, the Lancaster crash, and stories about the Brace (Wild River) Cabin.

Meliss Brown Clark, Steve and DeWitt Brown (C8) supplied information and photographs. They share our interest in preserving Pistol Creek Ranch and its heritage. The ashes of Steve's father, mother, and brother are buried in the cemetery. The ashes of DeWitt and Meliss' father and mother are also buried in the Ranch cemetery.

Chris and Dave Dewey (Ranch managers) supplied information on Pistol Creek Ranch history, cabin owners, Ranch history, and were great friends. Chris also supplied cabin photographs.

Harold Dougal (Backcountry pilot) supplied early Pistol Creek Ranch photographs. We are indebted to him for his knowledge of Ranch history, helping to build the Ranch, and alerting us to the poignant and incredible 1943 story about the survivors of the missing B-17 Bomber crew.

Mark Fermoile (C9) photocopied the Hornback survey and calculated Pistol Creek Ranch acreage.

Steve and Julie Kirby (C13) provided photographs and history of their cabin. Julie alerted us to the pictograph in Pistol Creek Rapids cave and Ross Geiling's cache outside the ranch gate. (C14) refers to Cabin 14.

Billie and Clint Gerlach (C5) supplied valuable Ranch history. Billie and Clint were friends of Harold Dougal and alerted us to Dougal's involvement and development of Pistol Creek Ranch. They also provided information for us to locate Bruce McKeighan, who was important in incorporating PC after the Hornbacks. Clint and Billie were also the first to come "on board" with their knowledge of Ranch history.

Barbara and Pat Hawn (Former Ranch managers) Pat supplied much Pistol Creek history and recorded the burial of Al Grant's ashes at the Ranch.

Lori (McKeighan) McKenna and Bruce McKeighan (C12) shared their knowledge of Ranch history. Lori also was a great help by providing a copy of the Ranch Survey and photographs.

Andy Patrick (Patrick Aviation) told us of his dad's ashes being placed at Pistol Creek and along the MF.

Mike Paulson (C18) shared with us his knowledge of Pistol Creek Ranch history.

Dave Pecora (C3) shared knowledge of Pistol Creek Ranch history and provided photographs. The ashes of Dave's father are buried in the cemetery.

John and Scott Schumacher (C21) shared photographs and knowledge of Pistol Creek Ranch history.

Lauren Slette (Former Ranch accountant) provided acreage of the Middlefork (non-private) property.

Blake Swanson (C2) did a stellar job of recording the chronology of the 2000 "monster firestorm".

Pat (Hornback) Vance (Daughter of the original Pistol Creek Ranch owner) supplied information about the original cabin owners and Ranch history.

Sonja and Bill Widgren (Ranch managers) supplied information and photos of their stay at the Ranch.

Diana Yupe (Shoshone Indian and historian at Indian Creek Guard Station) provided valuable history about early Shoshone life on the Middle Fork.

NOTE APS: I became aware of Pistol Creek Ranch in late 1970 when I and two friends took a float-trip down the Middle Fork. We stopped briefly at Pistol Creek Ranch where I knew Dr. Charles Fleming owned a cabin. At the time I was a member of a group of seven pathologists in Reno, Nevada. We serviced a number of hospitals that were too small to have a full-time pathologist. I made monthly visits to Portola, California's hospital, so their clinical lab could be accredited. At the hospital I met Drs. Charles Brown and Bill Bross, original owners of C8. Dr. Bross moved to Reno and became an ER doctor. We became friends and took a course to become certified in clock repair. Before buying C9 I flew to the Ranch with Dr. Bross and Carl Barlow, a Reno contractor to evaluate the cabin. In 1987, Dr. Fleming sold C9 to Arlene Sohn & Dr. Anton Sohn and Drs. Trudy Larson Dixon & Sherwood Dixon.

CONTRIBUTORS LISTED UNDER REMINISCENCES OF PISTOL CREEK, PAGE 66-77

1. BILLIE AND CLINT GERLACH
2. MELISS CLARK
3. DAVE PECORA
4. HAROLD DOUGAL
5. PAT HAWN
6. JOHN SCHUMACHER
7. SCOTT SCHUMACHER
8. RUSTY BRACE
9. MEREDITH BRACE
10. PATTY VANCE
11. BILL WIDGREN
12. BRUCE McKEIGHAN
13. LORI (McKEIGHAN) McKENNA
14. CHRIS DEWEY
15. MIKE PAULSON
16. ROBERT BOYD
17. STEVE KIRBY

ILLUSTRATIONS

1. BILL PAYNE AT PISTOL CREEK RAPIDS CAVE, 2013 1
2. PICTOGRAPH IN RAPIDS CAVE   PISTOL CREEK RAPIDS CAVE, 2013 1
3. NOTE: GRAYISH DISCOLORATION FROM ANCIENT FIRES. PICTOGRAPH IS THE SLIGHT PINK COLOR ON THE ROCK IN THE UPPER LEFTHAND. 3
4. AERIAL VIEW OF PISTOL CREEKAND THE MIDDLEFORK (Scott Schumacher) red locations added by aps 3
5. SAM HOPKINS AT HIS PISTOL CREEK CABIN, 1910 4
6. CUT WOOD AND SKETCH OF ARTIFACTS FOUND AT CABIN SITE, 2014 5
7. ARTIFACTS FOUND AT MITCHEL/MEYERS CABIN SITE, 2014 (shovel displayed to show size of foundation stones) 5
8. "LAND PATENT" FILED BY HOMESTEADER GEORGE RISLEY IN 1923, signed by president Calvin Coolidge. (D. Pecora) 6
9. ORIGINAL PLAT MAP FROM 1920, Central Idaho was the last part of the continental U.S. to be platted and homesteaded. (D. Pecora) 7
10. BILL WAYNE AT GEORGE RISLEY'S CABIN, 1950S 8
11. GCGS (D. Pecora) APS AT GCGS FOUNDATION RUINS 8
12. GARDEN CREEK GUARD STATION, 1940   USFS RUINS, BETWEEN PCR and PC 9
13. ED HUNTINGTON'S CABIN, Bill Thompson on a "frontier scooter", 1993 Cabin burned in the 2000 fire. Note the decades old, charred tree at the right. 9
14. SUSPENSION BRIDGE At Pistol Creek Rapids, 1971 (DeWitt Brown) (Later removed by U.S. Forest Service) 10
15. RICHARDSON'S PAINTING, 1960s (S. SCHUMACHER)   RUNWAY HANGER, 1962 11
16. KRISTIN S. FERMOILE (Ross' Cabin) 1992   ROSS GEILING, 1987 (Rusty Brace) 12
17. ROSS' CABIN at 45 CREEK, 1992   ROSS' CACHE at PC FLATS (before Pistol Ck.) 12
18. B. THOMPSON, APS, BILL SOHN, B. PAYNE,   ROSS GEILING CABIN SITE & B. MONTGOMERY AT ROSS' CABIN, 1992 12
19. ORIGINAL CABIN, C12, 1960 (Lori McKenna) 13
20. WAYNE'S ORIGINAL Lodge Facing the Middle Fork, 1960 13
21. JACKIE/RAY NIXON, C12, 1960   JEAN/DUANE TJUMSLOND 13
22. BILL WAYNE (LODGE AND SAWMILL), 1960   JULIE KIRBY, 2013 16
23. "WILD" BILL WATSON (Bea Gubler), 1962   BILL PECORA & HAROLD DOUGAL 16
24. CHARLES BROWN (Meliss Clark), 1960   MARYLEWIS BROWN (Meliss Clark), 1960 ...16
25. JEFF CROSBY, KIM GERLACH, LORI McKEIGHAN (Mckenna), and RUSTY BRACE (Lori McKenna) 1970s 17
26. DAVE ROBBINS AND BRUCE McKEIGHAN, DAN PLUNKETT, 1982 (Steve Kirby) 1950s ...17
27. DENNIS SMILANICH, KIM AND JOHN   SCOTT SCHUMACHER, 1970   DAVID PECORA, 1963 18
28. BILL BRACE AND JACK CONWAY, 1969 (At Indian Creek with Conway's DC-3) 18
29. H.L. TURNER (Pan Am Pilot)   BRUCE McKENNA 19
30. CLAY LACY (United Airlines pilot)   JOHN CONROY, CLAY LACY (Rusty Brace) 1960s 19
31. JOHN CONROY, PAUL COX, CLAY WOODS, BUZZ CHANEY, 1960 19
32. AND CLINT GERLACH, 1976 19
33. CARL BARLOW (Reno contractor)   BOB PAULSON AND PATTY McKENNA   AND BILL BROSS, 1987 21
34. DAVE AND SCOTT PECORA, 1970s   DON HEATER AND BILL BRACE, 1971 21

35.	BILL WATSON, UNKNOWN, FLOYD POSING, AND UNKNOWN, 1960s	22
36.	JOHN CONROY, 1960s	22
37.	LOGGING TO BUILD CABINS, 1960 BRIDGE AT LITTLE PISTOL CREEK, 1940	22
38.	(Scott Schumacher)	22
39.	UNKNOWN AND JOHN CONROY, 1960s	23
40.	LORI MCKEIGHAN (Lori McKenna), 1960s	23
41.	ROBT. BOYD, APS, KAREN BOYD, CHRIS DEWEY, & DAVID DEWEY, 2013	24
42.	BENJAMIN F. BROWN 1946-2003 BENJAMIN A. BROWN 1905-1998	24
43.	CHARLES W. BROWN MD 1906-1981 DAVID M. ROBBINS 1923-1979	24
44.	JOHN M. CONROY 1920-1979 MARYLEWIS S. BROWN 1914-1987	24
45.	BILL WAYNE 1903-1956 SCOTT PATRICK 1952-2011 (100 ft. uphill from cemetery) (1/2 Ashes at Side of Runway)	25
46.	ALFRED A. GRANT III 1920-1993 WILLIAM F. PECORA 1931-1999	25
47.	HELEN KIRKPATRICK BROWN 1919-2002 CHET LANCASTER 1925-1997	25
48.	DAN WAGNER 1951-1997 JOHN W. LANCASTER 1946-2000	25
49.	LOADING ENGINE FOR EXAMINATION IN CASCADE JOHN LANCASTER	28
50.	GENE TOOLE AT LANCASTER CRASH SITE, JULY 25, 2000	28
51.	LANCASTER CRASH SITE, JULY 25, 2000	28
52.	SIPs FOR REBUILDING, 2001	31
53.	SIPs FOR THE WALLS, 2001 SIPs FOR THE ROOF, 2001	31
54.	RANCH LODGE, 2023 (Not damaged by 2000 Fire)	32
55.	BARN CONSTRUCTION, 2001 BRACE'S (C16) NEW CABIN BURNED, 2017	32
56.	RANCH SAWMILL. 2001	32
57.	(C1) SWANSON (Original) (C2) SWANSON (Original)	33
58.	(C3) PECORA (built Post-fire) (C4) PATTERSON (built Post-fire)	33
59.	(C5) GERLACH (built Post-fire) (C6) GRAMMAR (not burnt)	33
60.	(C7) McCAW (built Post-fire) (C8) BROWN (built Post-fire)	33
61.	(C9) DIXON/Sohn (built Post-fire) (C10) BRACE (not burnt)	34
62.	(C13) MARKSTEIN/POWELL (built Post-fire) (C14) BOYD (built Post-fire)	34
63.	(C11) BROOKS (built Post-fire) (C12) TRaVIS/MARKSTEIN (not burnt)	34
64.	(C15) THORNTON (built Post-fire) (C16) BRACE (built Post-fire) (Destroyed by fire, 2017)	34
65.	SHOP/REPAIR FACILITY (built Post-fire)	35
66.	(C17) NARCHI/SPAULDING (built Post-fire) (C19) FOX (built Post-fire)	35
67.	(C21) SCHUMACHER, SHAHI/ERKEL (built Post-fire) MANAGER'S CABIN (built Post-fire)	35
68.	BARN/STAFF QUARTERS (built Post-fire) STAFF CABIN (Original)	35
69.	BILL & SONJA WIDGREN1., 1995 P. and B. HAWN2, 1996	36
70.	BILL WIDGREN, 2000 G. Toole, B. montgomery, B. Payne, H. brown, D. & C. DEWEY, 2014	36
71.	CHRIS and DAVE DEWEY3., 2013 P. HAWN, B. HAWN, O. BOLSTAD, J. FLANARY, R. HOGAN, PHIL SOHN, 1996	36
72.	SWAT TEAM AT PISTOL CREEK RANCH RUNWAY	37
73.	SWAT TEAM ON THE GROUND AT PISTOL CREEK RANCH	37
74.	SWAT TEAM IN A PISTOL CREEK CABIN AND WITH CHILD	37
75.	CHARLES FLEMING, 1970 BRADY AND KERRY SOHN, 2005	38
76.	KERRY SOHN, 2004 ROY HOGAN, 1994 CHRIS SOHN, 2010 (His first fish. It went back in the river)	38
77.	KRISTIN S. FERMOILE (Artillery Lake), 1992 G. TOOLE, B. MONTGOMERY, 1989	38
78.	BOLSTAD, HAWN, PHIL, HOGAN, 1989 PHIL, ERIC SOHN, 2011	39
79.	MATT SCHMITT, 2015 APS, KERRY, AND ERIC SOHN, 2016	39

#	Entry	Page
80.	MARK FERMOILE, 2007 BRADY SOHN, 2018	39
81.	PETER and PHIL SOHN, LANGHAM, HADLEY, JERRY, FRANK, and DOUG GLEASON, LIZ and ALEX SOHN, 2023	40
82.	ROB S., APS, CHRIS S., MONTY, BOB S., 1990 HOGAN, BOLSTAD, PHIL S., 1989	40
83.	P. HAWN, APS, CHRIS S., MONTY, TOOLE APS, PAYNE, THOMPSON, TOOLE, BILL S., at Buck Lake, 1990 MONTY at Artillery Lake, 1992	40
84.	KERRY SOHN, 2004 ISABELLA SOHN, 2015	44
85.	SIERRA SOHN, 2004 KRISTIN SOHN FERMOILE, 1992	44
86.	BILL PAYNE CROSSING THE MF, 2005 MIMI SOHN, DAVE DEWEY, 2010	44
87.	APS, BRADY SOHN, 2004 B. THOMPSON, FLINTLOCK RIFLE, 2004	45
88.	B. THOMPSON, B. PAYNE, APS, B. MONTGOMERY, 1999 BRADY SOHN, 2010	45
89.	BILL PAYNE, GENE TOOLE, BILL THOMPSON, MATT, SPENCER, and SWIMMING IN THE MIDDLEFORK, 1994 MAXIMILLIAN SCHMITT, 2023	45
90.	ARLENE, M. BROOKS, C. PAYNE, H. BROWN, G. TOOLE, APS, D. DEWEY, G. BROOKS, B. PAYNE, 2016 B. MONTGOMERY, B. PAYNE, 2014	46
91.	SACK RACE, JULY 4, 2013 DUCKIE RACE, JULY 4, 2016	46
92.	An Eagle too high to Identify	47
93.	CABIN NAME ON CANOE PADDLE (Made by APS)	48
94.	TABLE & NIGHT STAND (Made by APS from discarded drawer)	48
95.	BENCH & CARVED BEAR (by APS) P. SOHN & LAMP STAND, 2009	48
96.	D. DEWEY & B. PAYNE, 2013 NIGHT LAMPS (Eagle carved by APS) (Made by APS from burl wood)	48
97.	PATIO TABLE AND CHAIRS (Made by APS)	49
98.	FLINTLOCK RIFLE AND PISTOL (Made by APS from a kit)	49
99.	OUTDOOR TABLE, APS AND KERRY SOHN, 2012	49
100.	TABLE (Made by APS from discarded doors) 2010 TABLE (Made by APS) 2013	49
101.	TOOLE, THOMPSON, PAYNE, MONTGOMERY G. TOOLE (Watercolor painting) 2006	50
102.	MONTGOMERY (Flies for fishing) 1998 KERRY, ALEX, BRADY, PETER, SIERRA, COLLIN PERRY (Painted cork designs) 2013	50
103.	FOX (Trapped by Kerry Sohn) 2012 FRIENDLY SNAKE (Sierra Sohn) 2010	50
104.	BEAUTIFUL SANDWORT SPOTTED SAXIFRAGE	51
105.	INDIAN CREEK GOLDEN COLUMBINE (by Aps)	51
106.	BIG BALDY CRIMSON COLUMBINE (by APS)	51
107.	PETER SOHN ON SCOOTER, 2013	52
108.	DAVE DEWEY and PATRICK WARREN 245 yr.-old, 34" Diameter Ponderosa Pine, 2012	52
109.	COW CREEK TERRACES, 2013	56
110.	BUCK LAKE (From the trail), 1990 PISTOL CREEK VALLEY ('V' shaped fire-scorched valley formed by water-flow)	57
111.	ARTILLERY LAKE (From the ridge and at Lake level), 1992	58
112.	BOULDERS, PEBBLES, SAND AND CLAY RANCH TERRACE	58
113.	MIDDLE FORK EAST BANK (At Pistol Rapids) CHOCOLATE MIDDLE FORK	58
114.	EXPLODED GRANITE JASPER (MINERAL)	59
115.	FELDSPAR (MINERAL) HORNFELS (ROCK)	59
116.	PINK GRANITE (ROCK) DIORITE (ROCK)	60
117.	DIORITE PORPHYRY (ROCK) PATIO FLAGSTONE (ROCK)	60
118.	ELK CARBONATED HYDROXYAPATITE	61
119.	ROCK SLIDE PLUTONIC ROCK	61
120.	NODULE FRAGMENT (11 CM), 2015	63

INTRODUCTION
(It's a Real Good Bet, the Best is yet to Come)

Pistol Creek Ranch History commemorates one hundred and thirty-one years of settlers at Pistol Creek Ranch on the Middle Fork of the Salmon River. The first homesteader came to the area in 1892, but a cabin wasn't built until sometime around 1910. It is difficult to establish the exact date due to lack of records in Valley County where Pistol Creek Ranch is located.

The county was established in 1917, and the county recorder has few records of property transactions before that date. The Ranch was deeded as a homestead in 1920.

The demand for information on Pistol Creek Ranch's history caused us to "step up" and research the beginning of the property. This was also a good time to add geological features and expand the reader's "Pistol Creek Ranch Knowledge."

We will take you back to the earliest known Americans on the Middle Fork and show that pioneers and miners in the region supplanted the Native Americans. In the last half of the twentieth century, Pistol Creek became the retreat that we know now, a relief from the "hustle and bustle" of life in the "States."

Pistol Creek Ranch (PCR) known as Pistol Creek (PC) or simply The Ranch, has a rich history. In addition, the Middle Fork is a microcosm of American's western expansion. It was the site of Indian wars, frontier soldiers, mining, hunting and trapping, Chinese immigrants, homesteads, and ranching. It is our intent to give you background information to add to your "Pistol Creek Knowledge."

ONE THOUSAND YEARS AT PISTOL CREEK

THE EARLIEST INHABITANTS ARRIVED ca. 1000 AD

The earliest "known life" at Pistol Creek and on the Middle Fork is recorded by pictographs. There is no scientific evidence to accurately date these images, but archeologists estimate them to be approximately one thousand years old. The artists are unknown. It is thought that they used pigment from minerals, animals, or plants to make the images. Tuka-Deka known as Shoshone lived in Idaho's mountains for eons. The Indians who lived in the Salmon River mountains were known as Sheepeaters and Salmoneaters. Sheepeaters relied upon big horn sheep for sustenance and survival, but they also relied on salmon. They were peaceful until the intrusion of miners and settlers who moved into the area and challenged their homes and life-providing resources.

BILL PAYNE AT PISTOL CREEK RAPIDS CAVE, 2013

PICTOGRAPH IN RAPIDS CAVE

PISTOL CREEK RAPIDS CAVE, 2013

Settlers accused Sheepeaters of stealing horses near present day Cascade and killing two prospectors. To cap it off, five Chinese miners were killed near Loon Creek; however, the straw that broke the camel's back was the murder of two ranchers on the South Fork in May 1879. In fact, there is evidence that some of the crimes were perpetuated by whites dressed as Indians, but none of the murderers were identified. Nevertheless, the die was cast, Captain Reuben Bernard led Troop G of the First Cavalry toward Payette Lake to calm flaring tempers and subdue the Sheepeaters.

The first campaign against the Indians took place on the Salmon River. On August 20, 1879, Indians attacked troops at Soldier Bar on Big Creek. By the first of October, little had been accomplished. The soldiers, with a group of twenty Indian scouts, agreed to a surrender.

The Indians were forcibly removed as prisoners of war, and the invasion of settlers into the Middle Fork region continued. Names of rivers and areas persist from the military campaigns. Shoshone descendants of Sheepeaters exist and thrive in the Salmon River area.

PISTOL CREEK RAPIDS CAVE

The definition of a cave is obscure and may be misunderstood. A cave is "A natural cavity, recess, or a chamber, beneath the surface of the earth large enough for entry by a person." (R.L. Bates & J.A. Jackson; Dictionary of Geological Terms, New York, Doubleday, 3rd ed. 1984). Conclusion: Rapids Cave is a Cave.

Large and beautiful caves are common and abundant in eroded limestone topographical areas across the planet where sedimentary limestone rock dominates as ground rock. The chemical makeup of limestone bedrock is primarily soft calcium carbonate which dissolves easily when subjected to contact by acidic precipitation, weathering agents, and erosion.

On the other hand, because of their mineral content and hardness, bedrock formations of igneous and metamorphic rock are resistant to mechanical and chemical weathering agents. Caves are rarely found in bedrock areas that are igneous and/or metamorphic. A realistic theory of cave formation in igneous and metamorphic rock area is that movement of the bedrock surface material due internal earth crustal forces causes a void. Thus, earthquakes, volcanic eruptions, and landslides cause caves to form in igneous and metamorphic bedrock. This appears to be the cause of the cave at Pistol Rapids.

The Idaho Batholith area at Pistol Creek Ranch is formed of igneous and metamorphic rock that is covered by eroded mountain materials from glacial melt water. This burial process has been in action since the Wisconsin glacial period 17,000 years ago. The deposits of the glacial materials in the batholith area are recognized as terraces. Finding a cave in the Pistol Creek area is rare; however, one does exist near Pistol Creek Rapids. On the west side of the Middle Fork at the Rapids is a steep hill, and about 100 feet below the crest of the hill is a cave. It is not a splendid cave, but it is interesting because there are traces of paintings on the walls. The passage of time has caused fading of the colors, but one can see vague imagery. The cave is less than ten feet deep, but it provides good protection from the elements. There are no stalagmites, stalactites, or other structures often found in limestone caves.

Rapids cave was not formed by erosion or the dissolving of the rock mineral content. It was formed by a huge landslide on the eastern side of the Middle Fork eons ago. The materials involved came from high elevation. Just how or when this occurred is not discernible. The momentum of the landslide was great enough to force rock materials to pile high on the western side of the river.

Thrust and settlement of the rock materials left a hole, the rapids cave at Pistol Rapids. It is also possible that friable rocks were removed from the cave by early Indian inhabitants. In addition, eroded landslide materials have also caused erosion downstream. **Examine the pictograph but don't touch!**

NOTE: GRAYISH DISCOLORATION FROM ANCIENT FIRES. PICTOGRAPH IS THE SLIGHT PINK COLOR ON THE ROCK IN THE UPPER LEFTHAND.

AERIAL VIEW OF PISTOL CREEK AND THE MIDDLEFORK
(Scott Schumacher) red locations added by APS.

SAM HOPKINS AT PISTOL CREEK, 1910

Although there was mining activity along the Middle Fork after the Civil War, Sam Hopkins (Hoppins?) in 1892 was the one of the first settlers on the Middle Fork. In 1910 Hopkins moved from Loon Creek and built a cabin near Pistol Creek and the Middle Fork. Little is known about Sam, including the spelling of his last name. It is recorded that he had a Scottish father and a Cherokee mother (He had to be a good guy). Sam was a packer and hauled supplies for miners and ranchers. Sam's original cabin site, which was probably the first cabin built at the site of Pistol Creek Ranch, is unknown.

Much effort and time has been spent by APS and Dave Dewey trying to match the photograph of Hopkins at his cabin with 2017 ground findings at the Ranch. Unfortunately, we were unsuccessful. Following is a much-publicized photograph of the original cabin in 1910.

SAM HOPKINS AT HIS PISTOL CREEK CABIN, 1910

ELECK WATSON AT PISTOL CREEK, 1914

In 1912, Hopkins sold his cabin and surrounding land to Martha and Trapper Eleck Watson. Six years later, Watsons sold their 144.33 acres at Pistol Creek to George L. Risley and wife, who in 1920 filled homesteading rights with the county with Bill Mitchel and Charles Meyers, who lived across the river, as witnesses.

BILL MITCHEL/CHARLES MEYERS CABIN, 1920

Ranch Manager Dave Dewey (who knew of the site), Phil Sohn, Peter Sohn, and APS crossed the Middle Fork at the Ranch's swimming beach on August 8, 2012, and hiked up Cow Creek to locate the site of the Mitchel/Meyers cabin. Approximately 200 yards from the Middle Fork we identified a leveled area with numerous artifacts that indicated a previous cabin site. The area was within ten yards of Cow Creek and had nearby evidence of mining activity. (See following diagram and photographs of artifacts on page 7.)

CUT WOOD AND SKETCH OF ARTIFACTS FOUND AT CABIN SITE, 2014

ARTIFACTS FOUND AT MITCHEL/MEYERS CABIN SITE, 2014
(shovel displayed to show size of foundation stones)

FRANK CHURCH "RIVER OF NO RETURN" WILDERNESS CREATED

THE AREA IS A PROTECTED WILDERNESS AREA IN IDAHO. IT WAS CREATED IN 1980 BY THE UNITED STATES CONGRESS AND RENAMED IN 1984, THE FRANK CHURCH "RIVER OF NO RETURN" WILDERNESS AREA IN HONOR OF U.S. SENATOR FRANK CHURCH.

Hailey 026445   4—1023-R.

The United States of America,

To all to whom these presents shall come, Greeting:

WHEREAS, a Certificate of the Register of the Land Office at **Hailey, Idaho,** has been deposited in the General Land Office, whereby it appears that full payment has been made by the claimant **George L. Risley** according to the provisions of the Act of Congress of April 24, 1820, entitled "An Act making further provision for the sale of the Public Lands," and the acts supplemental thereto, for the **Lots one, two, five and six of Section eight in Township sixteen north of Range eleven east of the Boise Meridian, Idaho, containing one hundred forty-four acres and thirty-three hundredths of an acre,**

according to the Official Plat of the Survey of the said Land, returned to the GENERAL LAND OFFICE by the Surveyor-General;

NOW KNOW YE, That the UNITED STATES OF AMERICA, in consideration of the premises, and in conformity with the several Acts of Congress in such case made and provided, HAS GIVEN AND GRANTED, and by these presents DOES GIVE AND GRANT unto the said claimant the tract above described; TO HAVE AND TO HOLD the same, together with all the rights, privileges, immunities, and appurtenances, of whatsoever nature, thereunto belonging, unto the said claimant and to the heirs and assigns of the said claimant forever; subject to any vested and accrued water rights for mining, agricultural, manufacturing, or other purposes, and rights to ditches and reservoirs used in connection with such water rights, as may be recognized and acknowledged by the local customs, laws, and decisions of courts; and there is reserved from the lands hereby granted, a right of way thereon for ditches or canals constructed by the authority of the United States.

IN TESTIMONY WHEREOF, I, **Calvin Coolidge,** President of the United States of America, have caused these letters to be made Patent, and the Seal of the General Land Office to be hereunto affixed.

GIVEN under my hand, in the District of Columbia, the **SEVENTEENTH** day of **DECEMBER** in the year of our Lord one thousand nine hundred and **TWENTY-THREE** and of the Independence of the United States the one hundred and **FORTY-EIGHTH**

(SEAL)

By the President: *Calvin Coolidge*
By *Viola B. Pugh*, Secretary.
M. P. LeRoy, Recorder of the General Land Office.

RECORD OF PATENTS: Patent Number **926902**

"LAND PATENT" FILED BY HOMESTEADER GEORGE RISLEY IN 1921, signed by president Calvin Coolidge. (D. Pecora)

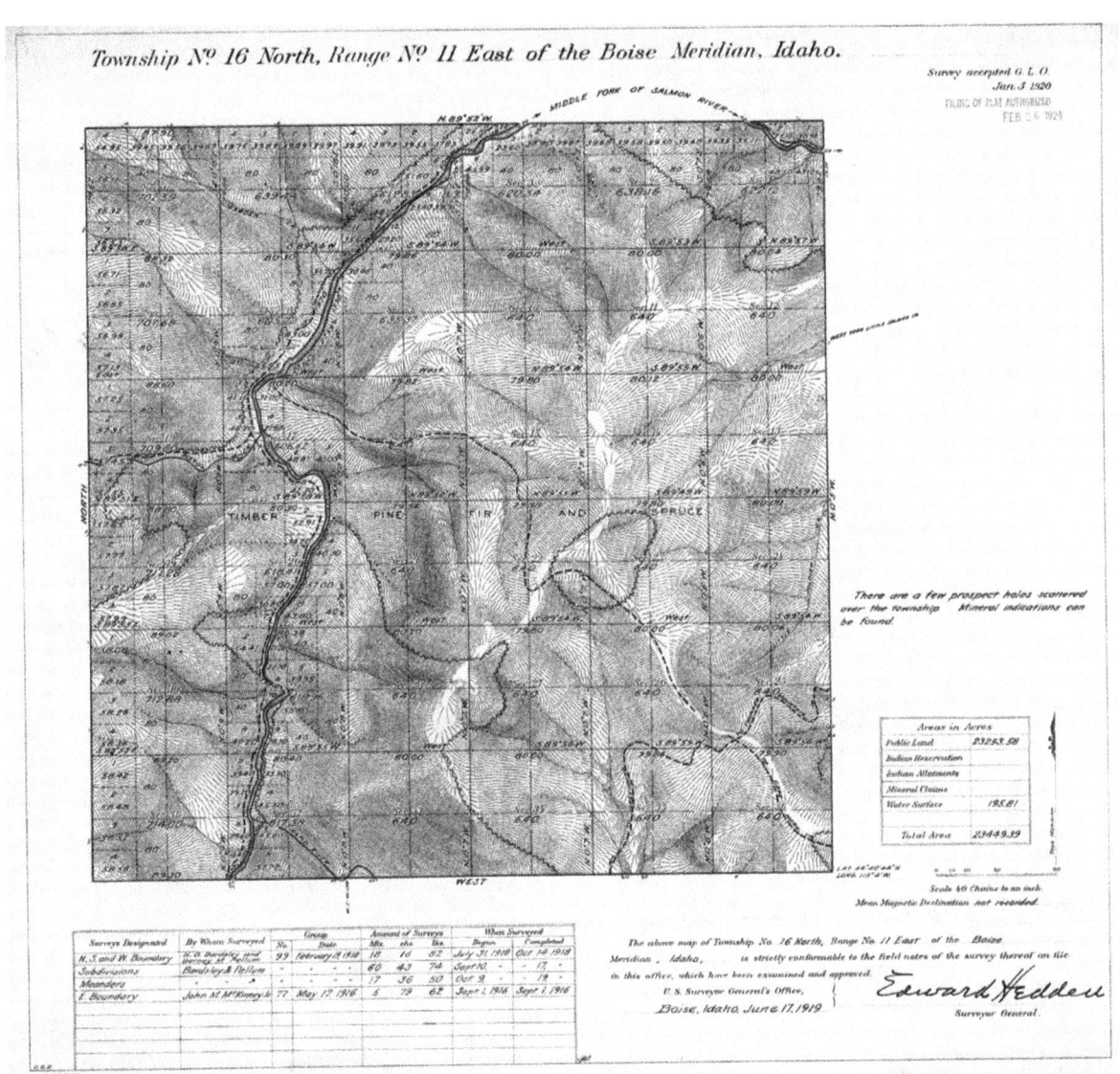

ORIGINAL PLAT MAP FROM 1920,
Central Idaho was the last part of the continental U.S. to be platted and homesteaded. (D. Pecora)

GEORGE RISLEY'S 144.33 ACRE HOMESTEAD, 1920

George Risley filed a homestead deed in 1920. He paid Watson $1.25 per acre ($30.82 per acre in 2023 when adjusted for inflation) for the Ranch with its cultivated thirty acres. Risley and his wife built two 16 feet × 20 feet cabins that were located across from the future runway. Bill Mitchel and Charles Meyers. lived across the river and witnessed Risley's homestead filing.

GEORGE RISLEY'S WIDOW SELLS RANCH TO BILL WAYNE, 1950s

After George Risley died, his elderly widow sold Risley Ranch to Bill Wayne in the early 1950s. According to Harold Dougal, Wayne paid around $4,500 for the property.

GEORGE RISLEY'S CABIN

BILL WAYNE AT GEORGE RISLEY'S CABIN, 1950S

GARDEN CREEK GUARD STATION (GCGS)

GCGS (D. Pecora)

APS AT GCGS FOUNDATION RUINS

The Garden Creek Guard Station where Harvey Wiegand spent three nights in 1943. (See page 15) It was moved to Indian Creek before Pistol Creek cabins were built in the 1960s. Foundation is north of Cabin 6.

Before Hornbacks established Pistol Creek Ranch, the US Forest Service (USFS) built several log cabins between Pistol Rapids and Garden Creek. The USFS has no records showing the location or numbers of cabins.

GARDEN CREEK GUARD STATION, 1940   USFS RUINS, BETWEEN PCR and PC

The US Forest Service (USFS) ruins are between the trail & the Middle Fork and are half way between the Ranch south gate and Pistol Creek.

GCGS AND ED HUNTINGTON'S CABIN AT FROG POND, 1938

ED HUNTINGTON'S CABIN, Bill Thompson on a "frontier scooter", 1993
Cabin burned in the 2000 fire. Note the decades old, charred tree at the right.

Ed Huntington and two miners filed a mining claim in 1938 and built the cabin at Frog Pond. Later, Huntington fled to South America after illegally manipulating stock.

SUSPENSION BRIDGE At Pistol Creek Rapids, 1971 (DeWitt Brown)
(Later removed by U.S. Forest Service)

PARACHUTE BAILOUT AT PISTOL CREEK, 1943

HARVEY WIEGAND, 1987: "On March 30, 1943, I flew over these mountains as a radioman on a training mission in a Boeing B-17 Heavy Bomber from Walla Walla, Washington. It was nighttime, and the weather was bad, we were running low on fuel, and one engine was giving us a lot of trouble. If we waited until it ran out of gas, we might not get out before it crashed. The pilot put us on autopilot, and we [nine crew members] jumped out into the dark."

After landing, I crossed the river in the junction of Rapid River and the Middle Fork and joined two other crewmembers. They struggled downstream in chest high snow until they found a rundown cabin (at Frog Pond) and spent the night "shivering." The next morning, they came to a wooden suspension bridge [above Pistol Rapids] and crossed the river and headed downstream. "I heard sounds like an animal or someone calling. The muffled sounds of voices got louder." "Hey, Harvey," I heard someone shout, "We're over here!" I looked up and saw two of the guys from our ill-fated B-17, my friends, standing in the snow on the other side of the river (Middle Fork)."

I forded the river, "I'll freeze for sure, if I don't get out of the middle of this river. Finally, and with much effort, I was able to get close enough to the other bank where my other buddies were. They helped me up the bank. We were all laughing now, and I was very much relieved. They told me that they had landed close to a log cabin [Garden Creek Guard Station]. The last one hundred yards more or less to the cabin seemed almost as far as the whole trip down the river. As we came into a small clearing, I saw the cabin ahead, a cabin made out of logs. It had a porch on the front of it and a small storage shed attached to the back."

HAROLD DOUGAL, 1987: "Later, the cabin was disassembled piece-by-piece, and the US Forest Service floated it downstream and reassembled it at the Indian Creek Guard Station. They took the porch off and installed sliding glass doors in the front. Then, they replaced the old handmade pine shingles with factory-made ones painted green." (The concrete foundation and steps remain near the Ranch on the Garden Creek side, north of Cabin C6. (See page 7)

MRS. HARVEY WIEGAND, 1987: "I'm so glad that you could fly us here. We weren't able to find anyone that really knew where the cabin had been until Scott Patrick told us he was sure that you would know."

HARVEY WIEGAND, 1987: "Well, after we all got back to the cabin, the boys told me they had found five sleeping bags up in the loft. They found a lot of food in the storage shed. There was only one can of meat, a canned ham that didn't last long and sure tasted good." There was a phone with a crank in the cabin, but they couldn't get it to work. One day when they were outside, they heard it ring and rushed inside to answer the phone. Mr. Hood at Thomas Creek and a woman were on the line talking. Harvey was too excited to wait for the conversation to continue, "Hey Lady! We have been stranded here for three nights. Can you help us?" After moments of silence, she replied that we must be flyers from the bomber that everyone is looking for. The next day, the five crewmembers hiked to the Indian Creek runway where Pilot Penn Stohr and Dick Johnson from McCall rescued them in a Fairchild Model 71 single engine plane equipped with skies. (Three more crewmembers were later found downriver near the Middle Fork and rescued.)

HAROLD DOUGAL, 1987: "I think it is only right that a moment of silence be given to the lad who did not survive and to District Forest Ranger Charlie J. Hamer and two U.S. Army Air Corps officers - Captain Bill Kelly and Copilot Arthur A. Crofts, who were killed in an airplane crash while on an air search mission looking for the downed B-17 crew."

ANTON SOHN: Harvey Wiegand's abandoned bomber crashed near Challis, Idaho, approximately fifty miles from Pistol Creek. The missing crewmember was never found. Wiegand was shot down on a European mission in August 1943 and spent the end of WWII in Stalag 17, a German prison camp where he was joined by two of his buddies from the ill-fated bomber from Walla Walla. They were U.S. Army Air Force WWII prisoners of war survivors and are genuine American heroes. Some information is added by the editor. This account of the rescue of the five crewmembers is abbreviated from Harold Dougal's book, *Adventures of an Idaho Mountain Pilot*. Wiegand told the story to Dougal in 1987 when Dougal flew him and his wife to Pistol Creek for an emotional view of the cabin where Harvey had spent three cold nights in 1943. The relocated cabin is at the upriver end of the Indian Creek Guard Station.

RICHARDSON'S PAINTING, 1960s (S. SCHUMACHER)

RUNWAY HANGER, 1962

Cessna's trade magazine, *Pennant*, recounted Hornback's development of PCR. Marvin, an Idaho rancher began flying in 1945 when he bought Sulphur Creek Ranch. In 1962 Marv bought his 17th airplane and 12th Cessna, a Skylark. Barbara Hornback also was a commercial pilot. Lori (McKeighan) McKenna has memories of the tree house to the left of the red hanger. The tree house also was a hangout for John and Scott Schumacher.

ROSS GEILING, 1987, and HIS CABIN at 45 CREEK

KRISTIN S. FERMOILE (Ross' Cabin) 1992

ROSS GEILING, 1987 (Rusty Brace)

ROSS' CABIN at 45 CREEK, 1992

ROSS' CACHE at PC FLATS (before Pistol Ck.)

B. THOMPSON, APS, BILL SOHN, B. PAYNE, & B. MONTGOMERY AT ROSS' CABIN, 1992

ROSS GEILING CABIN SITE

Ross visited the Ranch cabins at night to "borrow" food. In December 1987, Ray Arnold flew over his cabin and saw no smoke from the chimney. Pat and Barbara Hawn investigated and found Ross' frozen body with a self-inflicted gunshot wound from a rifle. He was pronounced dead in Cascade, January 9, 1988, 71 years old. The cabin burned in the 2000 fire.

June 13, 2013, Anton Sohn: Julie Kirby came by my cabin, and we went to Pistol Creek with Marykay Brooks to visit Ross Geiling's cache located just outside the Ranch gate on Pistol Creek flats before the creek. The cache is about 20 yards off the trail to the right when going to the Pistol Creek. It contained an ax, rusted pots/pans, and implements Ross had stored. See page 14.

RANCH'S ORIGINAL CABIN & LODGE, 1960

ORIGINAL CABIN, C12, 1960 (Lori McKenna)

WAYNE'S ORIGINAL Lodge Facing the Middle Fork, 1960

Jackie/Ray Nixon, C12, 1960

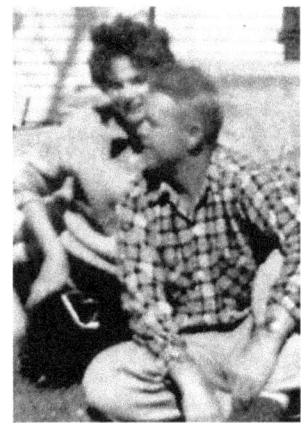

Jean/Duane Tjumslond

ADELAIDE AND BILL WAYNE, 1950s

In the 1950s, Adelaide and William E. Wayne built a two-story lodge at the ranch. (Photograph on page 15) Bill Wayne died May 12, 1956, in a backcountry plane accident, and Harold Dougal spread his ashes over the ranch. In 1957, Hornbacks bought the ranch.

SULPHUR CREEK TO PISTOL CREEK RANCH FLIGHT, 1950s

In the mid-1950s, Clay Lacy and John Conroy discovered Sulphur Creek Ranch while on training missions in National Guard F-86s. The Hornbacks, who were at Sulphur Creek introduced them to the area and led the way to Pistol Creek. Dave Robbins, McKeighans, and others followed to Pistol Creek Ranch.

HORNBACKS DEVELOP PISTOL CREEK RANCH, 1957

Marvin and Barbara Hornback had a vision of a "fly-in Ranch" at Pistol Creek. They caught the attention of Californian and western aviation pioneers. The group included pilots and aviators who served in World War II as well as commercial airline pilots. Hornbacks built a two-story lodge facing the river to replace the Wayne lodge. The Thornton cabin replaced the Hornbacks lodge. They named the property "Pistol Creek Ranch" and rented rooms in the lodge for hunters and fishermen. The only power at the time was propane, and the lodge was the focus of ranch activity with a canteen and laundry facilities.

Unfortunately, the lodge burned in 1958 while they were developing the Ranch. In 1959, they surveyed the property for forty-nine private lots. The Hornbacks built the landing strip and sold lots that were roughly one-half acre with a furnished cabin. They built cabins from 1959 into the 1960s by using money from the sales from one cabin to finance the next. Dewey Heater and son Don built their cabin near the site of the Riley cabin and supervised workers for the Hornbacks. Photo of Riley cabin is on page 95.

FIRST PISTOL CREEK RANCH CABINS, 1960

Doctors Charles ((Brownie) Brown and Willard (Bill) Bross from Portola, California, bought one of the first two-bedroom cabins. In 1959, they paid $9,500 for the cabin and $3,000 for the lot from Hornback for a total of $12,500. ($12,500 in 1960 = $128,847 in 2023) Gerlachs bought the last furnished cabin and lot for $18.000 with an additional $3,000 for Middlefork Ranch, Inc. stock.

BRUCE McKEIGHAN BUYS PISTOL CREEK RANCH AND INCORPORATES MIDDLEFORK RANCH, 1966

In October 1965, Marvin Hornback was killed in a homebuilt plane piloted by a friend at Strawberry Glen Airport near Boise. (Photograph of Marvin and Barbara Hornback is in the PCR Lodge hallway.) After Marvin's death, Bruce McKeighan bought the Ranch from Hornback's creditor in Boise for $70,000. In 1966, McKeighan established Middlefork Ranch, Inc. and offered each lot owner a share in the corporation for $3,000. Cabin owner Rex Lanham elected to not join the corporation. The Articles of Incorporation were signed August 22, 1966, and the ownership of the Ranch was turned over to the owners of the corporation. A board of directors manages the corporation and members provide a manager and assistants. The Ranch has rooms for members and ranch guests in the lodge. Horses and guides are provided for riding and hunting.

PISTOL CREEK RANCH, 144.33 ACRES, 1914

- 144.33 ACRES (Sold to Eleck Watson in 1914).
- About 14 ACRES (twenty-two private lots, including Grammar's future lot). The estimate is taken from the 1959 Hornback survey. A margin of error is due to the legibility of the copied survey (APS).
- About 139.75 ACRES (total PC area, including private lots, calculated from the 1959 Hornback survey). A margin of error is due to estimating the edge of the riverbank (Mark Fermoile, Aesthetic Engineering, Reno, Nevada).
- 130.9019 ACRES – Taxes paid on MF Corporation (non-private) property in 2012, but does not include twenty-two private lots of about 14 acres (Lauren Slette, Ranch Accountant).

RANCH SIZE – Using taxes on 130.9 acres, private lots at c. 14 acres, and estimated acreage from the 1959 Hornback survey, it is safe to say that the 2023 Ranch is 144+ acres. Due to the acreage listed in 1914 and its similarity to recent calculations, we can reliably state that the present size and shape of the Ranch dates to between 1910 and 1914 when Watson sold his 144.33-acre ranch. (Color added by APS)

BILL WAYNE (LODGE AND SAWMILL), 1960

JULIE KIRBY, 2013

"WILD" BILL WATSON (Bea Gubler), 1962

BILL PECORA & HAROLD DOUGAL

CHARLES BROWN (Meliss Clark), 1960

MARYLEWIS BROWN (Meliss Clark), 1960

JEFF CROSBY, KIM GERLACH, LORI McKEIGHAN (McKenna),
and RUSTY BRACE (Lori McKenna) 1970s

DAVE ROBBINS AND BRUCE McKEIGHAN,
1950s (Lori McKenna)

DAN PLUNKETT, 1982 (Steve Kirby)

DENNIS SMILANICH, KIM AND JOHN DAVID PECORA, 1963

JOHN AND SCOTT SCHUMACHER, 1970

BILL BRACE AND JACK CONWAY, 1969
(At Indian Creek with Conway's DC-3)

H.L. TURNER (Pan Am Pilot)

BRUCE McKENNA

CLAY LACY (United Airlines pilot)

JOHN CONROY, CLAY LACY (Rusty Brace) 1960s

JOHN CONROY, PAUL COX, CLAY WOODS, AND CLINT GERLACH, 1974

BUZZ CHANEY[1], 1960

1. PC Ranch manager, 1970s

ORIGINAL PISTOL CREEK RANCH CABIN OWNERS, 1960s

C1* BUZZ CHANEY HAD A MOTORCYCLE DEALERSHIP IN BOISE. FERRIS LIND SOLD THE LOT TO CHANEY, WHO BEGAN CONSTRUCTION IN 1969.

C2* BETTY AND DICK DeLONG; SALLY AND CHET LANCASTER; SMITH.

C3 JOAN AND BILL PECORA; WILMA AND DENNIS SMILANICH.

C4 VIRGINIA AND H. LANIER TURNER.

C5 BILLIE AND CLINT GERLACH BOUGHT THE LAST ORIGINAL CABIN IN 1967. THE OWNER BEFORE GERLACH WAS A RENO GROUP, BUT THEY SPENT NO APPRECIABLE TIME AT THE RANCH AND LEFT NO RECORDS.

C6* BEN GRAMMAR (PALO ALTO, CALIFORNIA) BOUGHT THE LAST CABIN THE RANCH BUILT. IT WAS BUILT IN 1980 WITH FLOWN-IN LOGS. BRUCE McKEIGHAN SUPERVISED.

C7 DEAN BROWN (HEAD BAKER FOR ALBERTSON'S) & JOE ALBERTSON (1 OF 3 MODEL CABINS). EARLY CABINS HAD SAWDUST INSULATION; LATER CABINS HAD NO INSULATION.

C8 MARY LEWIS AND CHARLES BROWN; WILLARD (BILL) BROSS. (CHARLES & WILLARD WERE FAMILY DOCTORS IN PORTOLA.) CALIFORNIA. CABIN BUILT IN 1960.

C9 SHIRLEY AND CHARLES (NEUROSURGEON) FLEMING (RENO, NEVADA). BUILT IN 1960.

C10* DR. JIM SCOTT (REDDING, CALIFORNIA). CABIN BUILT IN 1960.

C11 GENE BARTON (MINNESOTA). CABIN BUILT IN 1960.

C12* BRUCE McKEIGHAN, RAY NIXON, DURL BRADLEY, JEAN AND DUANE TJUMSLOND. CABIN BUILT IN 1960.

C13 DAN PLUNKETT (LAS VEGAS) BROUGHT IN A CREW TO BUILD THE CABIN.

C14 GEORGE DOVEL (BOISE, IDAHO) BUILT THE CABIN. HE FLEW IN WHITE ROCK FROM MARBLE CREEK FOR THE FIREPLACE.

C15 PISTOL CREEKL RANCH LODGE ON THE RIVER WAS REBUILT AFTER THE LODGE BURNED IN 1958. JACK THORNTON BOUGHT THE LODGE IN 1976.

C16 FERRIS LIND (BOISE, IDAHO) OWNED STINKER FILLING STATIONS IN IDAHO. (1 OF 3 CABINS).

C17 CARYL AND PAUL COX; 2 OTHER OWNERS. CABIN BUILT IN 1960. THE CABIN HAD 3 UNITS.

C18 ESTELLE AND DALE DOOLEY. BUILT IN 1967. (1 OF 3 MODEL CABINS).

C19 ALLEN PAULSON; LOIS AND CLAY LACY; DAVID ROBBINS. BUILT IN 1960.

C20 JACK MATTICH (CALIFORNIA). CABIN BUILT IN 1960.

C21 RICHARD RUSSELL (CALIFORNIA STATE SENATOR), RICHARDSON AND COLEMAN. CABIN BUILT IN 1960.

C22 GEORGE RISLEY'S LOG CABIN. LATER, BILL WAYNE'S, THEN IT BECAME DON HEATER'S NEW CABIN. THE HORNBACKS MOVED INTO THE HEATER CABIN AFTER THEIR CABIN BURNED. IT BECAME LANHAM'S.

INFORMATION ON THE FIRST PISTOL CREEK RANCH CABIN OWNERS IS FROM MELISS (BROWN) CLARK, DeWITT BROWN, DAVE PECORA, PATTY (HORNBACK) VANCE, McKEIGHANS, BRACES, SCHUMACHERS, AND CHRIS DEWEY. SOME DATA ON THE FIRST OWNERS IS CONFLICTING, AND WE COULD NOT VERIFY ALL THE INFORMATION.

* Five (C1, C2, C6, C10 and C12) of the original cabins remain after 2000 firestorm.

CARL BARLOW (Reno contractor) AND BILL BROSS, 1987

BOB PAULSON AND PATTY McKENNA

DAVE AND SCOTT PECORA, 1970s

DON HEATER AND BILL BRACE, 1969

BILL WATSON, UNKNOWN, FLOYD POSING, AND UNKNOWN, 1960s

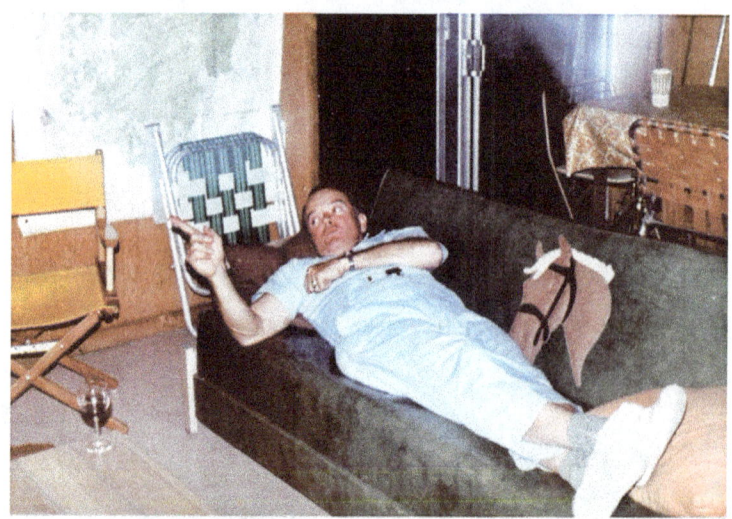

JOHN CONROY, 1960s

LOGGING TO BUILD CABINS, 1960

BRIDGE AT LITTLE PISTOL CREEK, 1940

(Scott Schumacher)

UNKNOWN AND JOHN CONROY, 1960s

LORI MCKEIGHAN (Lori McKenna), 1960s

PISTOL CREEK RANCH CEMETERY MARKERS

So live, that when thy Summons comes to join
The innumerable Caravan, that moves
To that Mysterious Realm, where each shall take
His Chamber in the Silent Halls of Death, William Cullen Bryant (1794-1878)

ROBT. BOYD, APS, KAREN BOYD, CHRIS DEWEY, & DAVID DEWEY, 2013

BENJAMIN F. BROWN 1946-2003

BENJAMIN A. BROWN 1905-1998

CHARLES W. BROWN MD 1906-1981

DAVID M. ROBBINS 1923-1979

JOHN M. CONROY 1920-1979

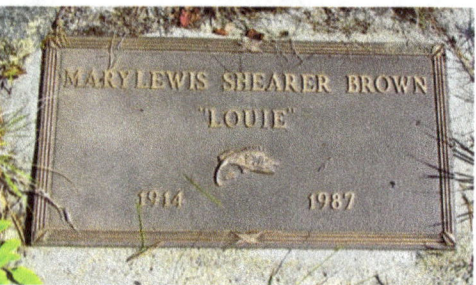
MARYLEWIS S. BROWN 1914-1987

BILL WAYNE 1903-1956
(100 ft. uphill from cemetery)

SCOTT PATRICK 1952-2011
(1/2 Ashes at Side of Runway)

ALFRED A. GRANT III 1920-1993

WILLIAM F. PECORA 1931-1999

HELEN KIRKPATRICK BROWN 1919-2002

CHET LANCASTER 1925-1997

DAN WAGNER 1951-1997

JOHN W. LANCASTER 1946-2000

PISTOL CREEK RANCH CEMETERY RECORDS, 2013

BENJAMIN ARVEL BROWN
13 Apr 1905-22 Dec 1998
MSgt. US Army Air Forces WWII, cabin owner, ashes in PCR cemetery

HELEN KIRKPATRICK BROWN
4 Nov 1919-13 Jun 2002
Benj. Arvel Brown's wife. Mother to Benj. F., Steven & Chas. Brown, cabin owner, ashes in PCR cemetery

BENJAMIN FRANCIS BROWN
27 Nov 1946-5 Aug 2003
Benj. Arvel & Helen K. Brown's son, father to Benj. P. Brown, cabin owner, ashes in PCR cemetery

CHARLES "BROWNIE" WILLIAM BROWN MD
1906-1981
Benj. A. Brown's brother, original cabin owner, ashes in PCR cemetery

MARYLEWIS SHEARER BROWN
1914-1987
Chas. W. Brown's wife, mother to De Witt Brown and Meliss Clark, ashes in PCR cemetery

JOHN "JACK" M. CONROY
1920 Dawn to Dusk 1979
Cabin owner, ashes in PCR cemetery

ALFRED A. GRANT, III
1920-1993
Cabin owner, ashes spread from the air by Clay Lacy and side of runway by Pat Hawn

CHESTER "CHET" FRANK LANCASTER
4 Jul 1925-24 Dec 1997
"Where I want to be". Original cabin owner, memorial only

JOHN WALTER LANCASTER
4 Mar 1946-25 Jul 2000
Ray Arnold Aviation pilot, died at PCR in aircraft engine failure, memorial only

SCOTT WILLIAM PATRICK
1952-2011
"Backcountry Pilot & Sportsman", Ashes at PCR & at locations along the Middle Fork

CAPTAIN WILLIAM F. PECORA
14 Dec 1931-14 Feb 1999
Father to David, original cabin owner, ashes in PCR cemetery & aircraft parking area

DAVID M. ROBBINS
1923-1997
Original cabin owner, ashes in PCR cemetery

DAN WAGNER
1951-1997,
Drowned at PC RAPIDS while on a river float trip, memorial only

BILL WAYNE
3 Mar 1903-14 May 1956
Plaque on a rock 100' uphill from cemetery, ashes spread from the air by H. Dougal

CLAY WOODS
19 Mar 1948-Sept 2003
'R.I.P'. Former Ranch manager 1973-77, ashes in Boise & in PCR cemetery by Robert Boyd

JOHN LANCASTER PLANE CRASH, JULY 25, 2000

The worst is yet to come,
   You won't know what it's all about
   Or where it's coming from.
Bill Widgren cried out, "Oh, Johnny!
   Black Ash Sunday"

APS' JOURNAL, JULY 25, 2000: As Bill Payne and I were painting the deck at cabin C9, a plane with Arnold's red stripes banked in front of the trees across the river from our cabin with its engine sputtering. It banked steeply and appeared to be on the way to the runway. All at once, we heard a loud sound like aluminum cans hitting concrete. We knew that the plane had gone down. Montgomery, Payne, Sohn, Thompson, Toole, and others ran toward the site, which was northwest of C9 and the runway. The crash site was in flames.

In the hurry to get to the crash, I accidently dropped my gloves and hat in the road behind the cabin. I later found them. By the time, we got to the crash, which was about one hundred and fifty yards from our cabin, the plane and surrounding vegetation were engulfed in flames. Ranch manager Bill Widgren, ranch hands, and individuals from the Brace Cabin rushed to the site.

From the breaking of two fifteen-inch diameter trees and height of the fall, it was apparent that no one survived.

Several individuals brought fire extinguishers and fire hoses, and we moved to extinguish the fire. Someone yelled, "Don't go near the plane." Why I did that I don't know. I probably needed to do something even though I saw that John didn't survive.

BILL WIDGREN cried out, "Oh, Johnny!"

Constantly, we heard popping, and we were concerned that a larger explosion was coming. Later, we found out that small cans were exploding. Bill Payne took over the fire hose.

The rest of the day was solemn. We had lost John, a good friend, and our pilot from three days earlier.

Later, it was determined that a faulty piston valve caused the engine to seize and stop.

IN MEMORY OF JOHN LANCASTER Letter from John's sister, Suzi Perry, January 24, 2012: "John Walter Lancaster was born March 4, 1946, in Huntington, Indiana. We, four children, were raised in the small town of South Whitley, Indiana, with a population, then and now, of about 1,500. Johnny was the third of four children and the 'only' boy.

"John's childhood was spent in the many woods surrounding S. Whitney with his dog Shag, hunting, trapping, and dreaming. He had a paper route from the age of ten and delivered 60+ papers on his bike and again, with Shag sometimes in several feet of snow.

"John headed to the West he loved, where he would spend the rest of his life. He graduated from the University of Montana to please our mother, because she wanted her children to have a college education, but he could have cared less. He returned to flying his true love. He was Evel Knievel's private pilot for two years (many great stories) and flew for American oil companies in Egypt. Aspen Airlines was another stop in his flying life. He was a Tanker Firefighter pilot battling forest fires before joining Arnold Aviation."

LOADING ENGINE FOR EXAMINATION IN CASCADE   JOHN LANCASTER

GENE TOOLE AT LANCASTER CRASH SITE, JULY 25, 2000

LANCASTER CRASH SITE, JULY 25, 2000

"BLACK ASH SUNDAY" AUGUST 27, 2000

AUGUST 10: Several hundred lightning-caused fires start in central Idaho. The fire that would reach the Ranch starts on Little Pistol Creek at Winchester Creek, nine miles west of the Ranch.

AUGUST 11: The USFS fire coordinator flies to the Ranch and tells Ranch Manager Bill Widgren that fire-fighting crews are needed elsewhere, and the USFS will not fight the Little Pistol Creek fire.

AUGUST 16: Five days later, the fire is growing and is eight miles from the Ranch. The Forest Service closes the Middle Fork.

AUGUST 21: Bill Widgren rides to the top of Big Baldy and reports the fire is at Trigger Creek, six miles from the Ranch.

AUGUST 22: All persons are advised to evacuate the area.

AUGUST 25: Fires are spotted on Garden Creek and Pistol Creek in areas less than two miles from the Ranch. The USFS coordinator again advises evacuation. Cabin owners are advised that the fire will reach the Ranch

AUGUST 26: Livestock is taken upriver to Seafoam Guard Station for evacuation.

AUGUST 27: The monster firestorm "consumes" seventeen cabins, melting aluminum roofs and glass windows. Only stone chimneys are left standing.

NOTE: The above is an edited version of Blake Swanson's written record of the fire.

APS COMMENT: The 2000 fire was not the first forest fire at Pistol Creek, and it will not be the last. A lightning-strike fire is Mother Nature's way of telling us, "I was here first." Prepare for a future fire by removing wood and vegetation from exterior cabin walls. Plant vegetation such as trimmed grass that is low risk for fires and will minimize the spread of a fire. All buildings should be constructed with steel or fireproof roofing. Remove dead vegetation and pine needles surrounding buildings. After the 2000 fire, the pond (reservoir) next to the runway was added to provide water for future fire protection.

POST-FIRE CABINS/LOT OWNERS, 2000

C1:	SWANSON	NOT BURNT
C2:	SWANSON	NOT BURNT
C3:	PECORA	TO BE REBUILT
C4:	PATTERSON	TO BE REBUILT
C5:	GERLACH	TO BE REBUILT
C6:	GRAMMAR AND KEMPF	NOT BURNT
C7:	McCAW	TO BE REBUILT
C8:	BROWN, CLARK AND WEAVER	TO BE REBUILT
C9:	DIXON AND SOHN	TO BE REBUILT
C10:	BRACE, HOLLISTER AND HASKEL	NOT BURNT
C11:	BROOKS	TO BE REBUILT
C12:	MIDDLETON	NOT BURNT
C13:	MARKSTEIN, KIRBY AND PRICE	TO BE REBUILT
C14:	BOYD	TO BE REBUILT
C15:	THORNTON	TO BE REBUILT
C16:	BAHAN	TO BE REBUILT
C17:	ZEMITIS	NOT REBUILT
C18:	PAULSON	NOT REBUILT
C19:	CHRISTENSEN	TO BE REBUILT
C20:	LACY	NOT REBUILT

C21: SCHUMACHER, SHAHIN, ERKEL AND COLLARD	TO BE REBUILT
RANCH LODGE	NOT BURNT
STAFF CABIN	NOT BURNT
BARN / SHOP BUILDINGS	TO BE REBUILT
MANAGER'S CABIN	NEW

Note: fourteen of seventeen destroyed cabins were rebuilt within three years after the fire.

REBUILDING CABINS, 2001

The first six cabins (C3, 8, 9, 11, 14 & 21) constructed in 2001 after the fire were built with Structural Insulated Panels (SIPs) that are factory-made particleboard walls, ceilings, and floors with foam board-insulating cores. The walls are pre-wired and plumbed.

SIP is another way to say prefabrication. This concept dates to the early 1900s when Sears sold prebuilt homes from catalogues. After WWII, National Homes Corp. and others shipped factory-built homes that were less expensive than conventionally constructed buildings.

CABINS CONSTRUCTED WITH SIPs, 2001

The prebuilt panels have several advantages. They are material efficient because lumber scraps are reused. In addition, the buildings are built in a few days, whereas conventionally built homes take weeks and even months.

An added advantage is that the panels are more airtight and heat/cold-efficient than rolled fiberglass insulation.

In contrast to most of the original cabins, the new cabins had complete peripheral foundations. Also, the first septic tanks of concrete blocks without leech fields were replaced with regulation composite septic tanks. To protect the river, the new septic tanks and leech fields were constructed to the west away from the river.

The USFS made available over $100,000 to each owner with a destroyed cabin to compensate for its lack of fighting the fire. To receive the money the owners agreed to certain conditions, including moving the cabin back a few feet from the river back. Some cabin owners rejected the stipulations and the compensation. A new hydroelectric generator was installed for safe water drinking. In addition to drinking waterlines, the Ranch installed irrigation and waterlines to each cabin for fire protection.

To further protect the Ranch and cabins from future fires, bushes and trees were removed from cabin walls, and grass with irrigation was planted. Furthermore, steel and fire-retardant shingles were installed.

A new manager's cabin, barn, workshop, garage, and fuel storage facility were constructed with metal roofing. After the fire, the sawmill was relocated near the barn that was being rebuilt. In addition to Ranch projects, lumber was milled for new cabins. Although the initial cabins were built using SIPs, it was necessary to supply timbers for beams, posts, porches, and storage buildings where insulation was not required.

After the barn, garage, and storage shed were finished, the sawmill was moved to accommodate the new horse barn.

SIPs TO REBUILD CABINS

SIPs FOR REBUILDING, 2001

SIPs FOR THE WALLS, 2001

SIPs FOR THE ROOF, 2001

REBUILDING THE RANCH

RANCH LODGE, 2023 (Not damaged by 2000 Fire)

BARN CONSTRUCTION, 2001

BRACE'S (C16) NEW CABIN BURNED, 2015

RANCH SAWMILL. 2001

POST-FIRE CABINS

(C1) SWANSON (Original)   (C2) SWANSON (Original)

(C3) PECORA (built Post-fire)   (C4) PATTERSON (built Post-fire)

(C5) GERLACH (built Post-fire)   (C6) GRAMMAR (not burnt)

(C7) McCAW (built Post-fire)   (C8) BROWN (built Post-fire)

(C9) DIXON/SOHN (built Post-fire)

(C10) BRACE (not burnt)

(C13) MARKSTEIN/POWELL

(C14) BOYD (built Post-fire)

(C11) BROOKS (built Post-fire)

(C12) TRaVIS/MARKSTEIN (not burnt)

(C15) THORNTON (built Post-fire)

(C16) BRACE (built Post-fire)
(Destroyed by fire, 2017)

SHOP/REPAIR FACILITY (built Post-fire)

(C17) NARCHI/SPAULDING (built Post-fire)

(C19) FOX (built Post-fire)

(C21) SCHUMACHER, SHAHI/ERKEL (built Post-fire)

MANAGER'S CABIN (built Post-fire)

BARN/STAFF QUARTERS (built Post-fire)

STAFF CABIN (Original)

Note: (C18) and (C20) were not rebuilt after the fire.

PISTOL CREEK RANCH MANAGERS

BILL & SONJA WIDGREN[1], 1995

P. and B. HAWN[2], 1996

BILL WIDGREN, 2000

G. TOOLE, B. MONTGOMERY, B. Payne,
H. BROWN, D. & C. DEWEY, 2014

CHRIS and DAVE DEWEY[3], 2013

P. HAWN, B. HAWN, O. BOLSTAD,
J. FLANARY, R. HOGAN, PHIL SOHN, 1996

1. Ranch manager 1997-2002. 2. Ranch manager 1985-1997 & intermittently. Gene & Mike Swenk (?) were mangers before Hawn. 3. Ranch manager 2005-2019. *Ranch manager Paul & Jessie Miller 2023.

Pat Hawn: "There was nobody on the ranch the past 1985 winter, and the property had been vandalized. Somebody had gone thru the cabins and took items of value and stockpiled them at Bruce McCall's cabin. The thieves flew the stuff out or floated it out by boat. When I arrived there was still a lot of stuff left in Bruce's cabin The thieves couldn't take it or didn't want what was left."

APS: Ross Geiling is the most likely suspect.

SWAT TEAM AT PISTOL CREEK, 2013

In 2013 a girl was kidnapped in Southern California and was driven to Idaho to be taken to the mountains. Authorities thought Pistol Creek was the destination, and a SWAT team arrived to search for the kidnapper and girl. They were later found at a campsite near Cascade, Idaho, where the kidnapper was shot and killed.

SWAT TEAM AT PISTOL CREEK RANCH RUNWAY

SWAT TEAM ON THE GROUND AT PISTOL CREEK RANCH

SWAT TEAM IN A PISTOL CREEK CABIN AND WITH CHILD

FISHING
(Food For The Hungry)

CHARLES FLEMING, 1970   BRADY AND KERRY SOHN, 2005

KERRY SOHN, 2004   ROY HOGAN, 1994   CHRIS SOHN, 2010
(His first fish. It went back in the river)

KRISTIN S. FERMOILE (Artillery Lake), 1992   G. TOOLE, B. MONTGOMERY, 1989

HUNTING
(It Doesn't Get Any Better Than This)

BOLSTAD, HAWN, PHIL, HOGAN, 1989

PHIL, ERIC SOHN, 2011

MATT SCHMITT, 2015

APS, KERRY, AND ERIC SOHN, 2016

MARK FERMOILE, 2007

BRADY SOHN, 2018

CAMPING
(Under the Stars)

PETER and PHIL SOHN, LANGHAM, HADLEY, JERRY, FRANK,
and DOUG GLEASON, LIZ and ALEX SOHN, 2023

ROB S., APS, CHRIS S., MONTY, BOB S., 1990

HOGAN, BOLSTAD, PHIL S., 1989

P. HAWN, APS, CHRIS S., MONTY, TOOLE
at Buck Lake, 1990

APS, PAYNE, THOMPSON, TOOLE, BILL S.,
MONTY at Artillery Lake, 1992

"BALLADS" AND "BLUES"

"PETRO BALLAD"[2]
By Robert E. Montgomery

When strolling thru the out-of-doors,
If a pretty rock you see.
Pick it up and admire,
The royal minerals that be.

If the sample you hold be igneous,
It will be granite, by all odds.
"Cause granite is land's most abundant,
Akin to ancient Gods.

If the stone you hold has smallest grains,
Alike sized, all near the same,
Sedimentary is what your holding.
Call it by that name.

Perhaps you may spot some symmetry,
Seashells or the like.
Limestone is the type you hold,
Remains of aqueous life.

The odds are small the rock you hold,
Will be metamorphic, as such.
Not appearing as originally formed,
It has been altered- Much.

Not knowing the name of the rock you hold?
Do not feel at a loss.
The real thrill in picking up a rock
Is giving it a toss.

---

2. Petra (Greek) = of stone, petrography = writing about stones.

"BALLAD OF PISTOL CREEK"
By Robert E. Montgomery

Going up Pistol Creek, going on a run,
Going up Pistol Creek to have a little fun.
Andy's got a cabin on Pistol Creek Ranch,
Arnold flies you in when he gets a chance.
Middle Fork Salmon is cold and clear,
Roar of Rapids in your ear.
No women to nag, no phones to hear,
Drag on a stogie, sip some cheer.
Sohn and Monty, Thompson and Toole,
Couldn't get Cox to ride Mollie, the mule.
Payne ties flies for the trout to eat,
Rainbows, cut-throats jumping at your feet.
Ponderosa, Jeffery pines reach high,
Tower into the deep blue sky.
Fire Brush red and Pussy Toes white,
Wild flowers cover the mountain site.
Ride up the mountain on the pack-horse trail,
Camp out in the sun and hail.
Atop the world where you touch a star,
Awe in wonder where you are.
Coyotes, hawks, an' bear, an' beaver,
Hummingbirds dart from feeder to feeder.
The elk are shy, but the deer come near,
Graze by the cabin without fear.
Thunder storms roll with bolts in the sky,
Trees ablaze when the brush is dry.
Smoke filled valleys and fires of red,
Thunder booming in your head.
The Idaho Mountains and the deep blue sky,
Gonna love Pistol Creek til the day I die,
Gonna love this place til the day I die,
Til the day I die.
We took a trip up to Pistol Creek.
We stayed for 'most a week.
Yeah, we took a trip to Pistol Creek
And stayed for a week.
We had a blast.

"PISTOL CREEK BLUES"
By Owen C. Bolstad

Don't fear—no chicken, pork or beer!
Andy made corned beef and cabbage.
His dinner was quite a winner.
Corned beef and cabbage for dinner
Was quite a winner.
Cabbage causes Intestinal gas.
Roy let a triple-flutter-blast.
Saw an UFO Thursday night,
Which caused quite a fright.
That UFO on Thursday night
Gave me quite a fright.
Turned out to be a Boeing jet,
But I see a Flying Saucer yet!
We planted seven little trees
And left them there to freeze.
We planted seven little trees
And left them there to freeze.
They'll grow up so big and tall.
They will burn, and then they fall.
The new cabin is really great.
I washed paper plates!
Yes, it was great,
Even when washing paper plates.
We will always remember that date,
And we will celebrate.
A man grows old, and then he dies,
But don't feel bad, and please don't cry.
SO HOIST A GLASS
And let's have fun,
Let's drink a toast to everyone.
Just lift that glass.
And have some fun.
Here's a toast to EVERYONE.
Hope you find "Crepitation Contest"!
A Song about a farting contest.

HORSEBACK RIDING
(Getting the Lead Out)

KERRY SOHN, 2004

ISABELLA SOHN, 2015

SIERRA SOHN, 2004

KRISTIN SOHN FERMOILE, 1992

BILL PAYNE CROSSING THE MF, 2005

MIMI SOHN,[3] DAVE DEWEY, 2010

3. Mimi's horse decided to take a bath in Frog Pond and forgot to tell Mimi.

GUN RANGE
(Never Miss a Shot)

APS, BRADY SOHN, 2004

B. THOMPSON, FLINTLOCK RIFLE, 2004

B. THOMPSON, B. PAYNE, APS, B. MONTGOMERY, 1999

BRADY SOHN, 2010

WATER SPORTS AND HIKING

BILL PAYNE, GENE TOOLE, BILL THOMPSON, SWIMMING IN THE MIDDLEFORK, 1994

MATT, SPENCER, AND MAXIMILLIAN SCHMITT, 2023

SOCIAL ACTIVITIES

(Love Your Neighbor As Yourself — *BIBLE*: Mark 12:31)

GENTLEMEN'S SERENITY CLUB OF PISTOL CREEK, 2004

ARLENE, M. BROOKS, C. PAYNE,
D. DEWEY, G. BROOKS,[4] B. PAYNE, 2016

H. BROWN, G. TOOLE, APS,
B. MONTGOMERY, B. PAYNE, 2014

SACK RACE, JULY 4, 2013

DUCKIE RACE, JULY 4, 2016

4. Gary Brooks killed in horse accident: Ashes—PCR cemetery; Forest Serv. Foundation; & Garden Creek.

BIRDS AT THE RANCH
Hast Thou Named All The Birds Without A Gun?
Ralph Waldo Emerson (1803-1882)

An Eagle too high to Identify

HANDCRAFTS

CABIN NAME ON CANOE PADDLE (Made by APS)

TABLE & NIGHT STAND (Made by APS from discarded drawer)

BENCH & CARVED BEAR (by APS)   P. SOHN & LAMP STAND, 2009

D. DEWEY & B. PAYNE, 2013   NIGHT LAMPS
(Eagle carved by APS)   (Made by APS from burl wood)

PATIO TABLE AND CHAIRS (Made by APS)

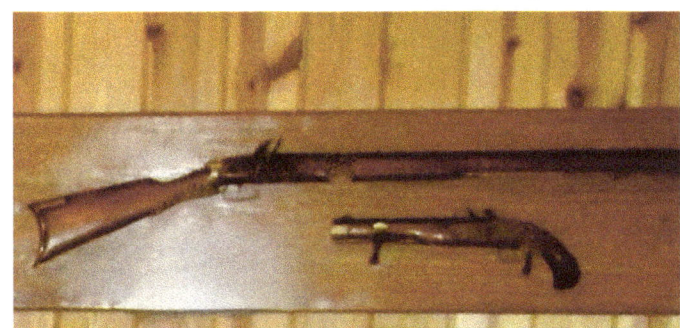
FLINTLOCK RIFLE AND PISTOL (Made by APS from a kit)

OUTDOOR TABLE, APS AND KERRY SOHN, 2012

TABLE (Made by APS from discarded doors) 2010   TABLE (Made by APS) 2013

CRAFTS
(Building a Better World)

TOOLE, THOMPSON, PAYNE, MONTGOMERY[5]

G. TOOLE (Watercolor painting) 2006

MONTGOMERY (Flies for fishing) 1998

KERRY, ALEX, BRADY, PETER, SIERRA, COLLIN PERRY (Painted cork designs) 2013

WILDLIFE
(Can You Find More?)

FOX (Trapped by Kerry Sohn)[6] 2012

FRIENDLY SNAKE (Sierra Sohn) 2010

5. Art class by APS
6. Coyotes and bears also visit the ranch.

AREA WILDFLOWERS
(Loved The Wood-Rose and Left It On Its Stalk—Ralph Waldo Emerson)

BEAUTIFUL SANDWORT

SPOTTED SAXIFRAGE

INDIAN CREEK GOLDEN COLUMBINE (by Aps)

BIG BALDY CRIMSON COLUMBINE (by APS)

PETER SOHN ON SCOOTER, 2013

DAVE DEWEY AND PATRICK WARREN
245 yr.-old, 34" Diameter Ponderosa Pine, 2012

PISTOL CREEK RANCH CABIN OWNERS, 2023

C1 & 2: BLAKE, TAD, KATHY, SALLY AND WILLIAM SWANSON
C3: WILMA, SCOTT AND DAVE PECORA; SEAR BLACK
C4: ALAN, CONNIE, DONNA AND BILL PATTERSON
C5: CLINT, BILLIE, AND CLINT III GERLACH
C6: JOAN AND BEN GRAMMAR
C7: JOLENE AND BRUCE McCAW
C8: CHARLES, STEVEN, DEWITT, PATRICK; BENJ. BROWN FAMILY TRUST; MELISS AND CHUCK CLARK; NANCY AND SCOTT WEAVER
C9: ARLENE AND ANTON SOHN; STEPHANIE AND MATT SCHMITT
C10: WILLIAM, MARY LOU, BILL AND ROBERT L. BRACE
C11: COOKIE KENT
C12: ELLEN AND RICK MIDDLETON
C13: HAYDEN, RAENEL, STEVEN MARKSTEIN; JULIE AND STEVE KIRBY
C14: KAREN AND ROBERT BOYD
C15: JACK, GAIL AND JOHN THORNTON
C16: MARIA FERNANDA ELOSU AND ROBERT L. BRACE
C17: QUINN SPALDING; ELSHA AND MICHAEL NARACHI
C18: BARRY BROWN
C19: MARGHERIA AND MATTHEW FOX
C20: LLC c/o NANCY LUKACS
C21: SCOTT SCHUMACHER; KRISTINE AND JOHN SCHUMACHER; JULIA AND ANSIA SHAHIN; DONNA AND AL ERKEL

MIDDLE FORK GEOLOGY
By Robert E. Montgomery

OREGON METEORITE STRIKE

A huge meteorite struck the Earth approximately seventeen million years ago in the southeast corner of Oregon. The meteor struck the earth with such impact that it fractured the crust allowing molten material from the mantel to penetrate the crust. This location is where a geological phenomenon today is known as the "Yellowstone Hotspot." The hotspot is a stationary deposit of molten magma in the earth's mantle below the earth's crust. The North American Tectonic Plate moved westward, colliding with the Pacific Tectonic Plate, but the hotspot remains in place.

SNAKE RIVER PLAIN

The Snake River Plain enters Idaho in the southwest corner of the state and ends in Yellowstone National Park. The route is a scar on the Earth's surface revealing the path of the North American Continent Tectonic Plate as Spokane passed over the Hotspot, which is the source of the volcanic activity that has been active across southern Idaho for nearly 17 million years. Scars of the volcanic activity are seen along the Snake River Plain in features such as lava beds, Twin Falls, American Falls, Craters of the Moon, and Idaho Falls.

Tectonic plate activity has created the Seven Devil's Complex on Idaho's western border. In this complex are the remains of several volcanic islands, which once existed in the Pacific Ocean. These islands were scraped off onto the No. Amer. Plate when the Pacific Plate was sub ducted beneath the No. Amer. Plate.

The Snake River has eroded through volcanic bedrock to form Hells Canyon. The river canyon extends north into the Columbian Plateau volcanic deposits where it joins the Columbia River complex.

LAKE MISSOULA

Lake Missoula was formed many times during the Pleistocene epoch when huge ice dams dammed a major drainage route for melt waters from glaciers. These floods happened at least forty-one times during the Pleistocene ice age that ended approximately 20,000 years ago. When the ice dam melted, 500 cubic miles of melted ice trapped behind the dam was rapidly released. A wall of water estimated by geologists to have been 2,000 feet high surged out of the lake causing catastrophic damage to everything in its path. These were the most destructive flood events in geologic history and were unique to Idaho.

Elusive evidence of these floods stumped field geologists for decades. Along a line drawn from Missoula, Montana to Spokane, Washington are hills of deposited sand and gravel that resemble waves on the ocean. This area is called Channeled Scablands of Washington. Not until geologists recognized these wavy hills as ripple marks, which are usually formed in streambeds, which the Missoula Lake enigma revealed.

Idaho is a geological wonderland. Her surface reveals almost all the results of building and destruction that geologic forces offer. This includes land twisted and bent out of shape by tectonic forces, mountains broken apart by faults, earthquakes, and crushed beneath tons of glacial ice. Sections of her crustal surface are uplifted by hot magma deposited in the deep voids vacated by lava from volcanic eruptions. U-shaped valleys are formed by glaciers. V-shaped valleys are formed by running waters, and lakes are formed by moraine deposits, etcetera. Rapids Cave is the result of these changes.

IDAHO BATHOLITH

Idaho's Bitterroot Mountains were formed by the exposure of an eastern section of the Bitterroot Batholith, which is the northern portion of the Idaho Batholith. The southern portion of the Idaho batholith where PC is located is known as the Atlanta batholith.

IDAHO BATHOLITH MATERIALS

The Idaho batholith is a vast expanse of granite that spreads throughout central Idaho. It is estimated that deposition of the magma occurred at a depth of over ten miles beneath the earth's crust during three different geologic periods ranging from 60 to 150 million years ago.

The rock families present in the Idaho batholith are igneous and metamorphic with salt and pepper rocks thrown in between.

IGNEOUS ROCK

Igneous rock is formed from molten minerals originating in the earth's mantle. This molten, mineral mass when deposited is called a pluton.

A pluton contains a vast, abundant, assortment of molten minerals. The pluton minerals, when they solidified as the Idaho batholith, became mostly gray granite. Keep in mind that "rocks" are a mixture of minerals in different concentrations. The color of any rock will depend upon its mineral content.

Igneous rock names are determined by the mineral content. This makes it difficult for a rock collector to properly identify and name igneous rocks.

SALT AND PEPPER ROCKS

"Salt and pepper rocks" are common white and black rocks around PCR. Only by use of lab analysis can each be tagged with its proper, scientific name.

METAMORPHIC ROCK

The other family of rock present the batholith is metamorphic. When the pluton intrusions of the Jurassic occurred, the basement rock of the crust was altered by the temperature and pressure caused by invasive pluton. There are three types of significant metamorphism.

1. Contact: when the new molten material meets the existing rock.
2. Regional: when temperature and pressure of the pluton is near enough to cause alteration.
3. Hydrothermal: when water from the aquifer, under pressure and super-heated by the intrusive pluton, alters existing rock.

All three of these processes were active in rock formation in the Idaho batholiths. Solid rock deposits in the Idaho batholith were metamorphosed when the younger Cretaceous magma intruded the older Jurassic batholith. This is also true when Eocene plutons intruded the older Jurassic and Cretaceous beds.

GEOLOGY DEFINITIONS

BATHOLITH: A large emplacement of igneous intrusive (also called plutonic) rock that forms from cooled magma that extends ten miles deep in the earth.

LATE CRETACEOUS: A geologic period and system lasting from 65 to 80 million years ago (MYA)

EOCENE: A geologic period and system of 60 MYA. IGNEOUS; One of the three main rock types, the others being sedimentary and metamorphic rock. Igneous rock is formed through the cooling and solidification of magma or lava.

LATE JURASSIC: A geologic period and system lasting from 135 to 150 MYA.

MAGMA: A mixture of molten or semi-molten rock.

METAMORPHIC: Means change in form, such as a rock that is changed by heat and/or pressure.

PLEOSTOCENE: A geologic period and system lasting from c. 2.588 MYA to 11,700 MYA.

PLUTON: Intrusive igneous rock (called plutonic rock) that is crystallized from magma that is slowly cooling below the surface of the earth.

SCABLANDS: Flat elevated land surfaces that are deeply scared by glacial flow/melted ice.

TECTONIC: relates to the building of the earth's crust.

GETTING ORIENTED AT PISTOL CREEK RANCH

For direction orientation at PR use the landing strip for reference. The strip runs southwest to northeast pointing nearly forty-five degrees from true north. Sighting along the airstrip from the takeoff pad to the northeast, to the far right will be the MF flowing northeast at the base of Cow Creek Mountain.

Notice how flat the land is where the landing strip is located. This is unusual in mountainous country. This terrace, once a river bottom, allowed construction on Ranch property, and it extends to the mountain on the west. This was the west bank of the MF when the melt waters from glaciers flowed through the Ranch's location. Behind you, south of the ranch on the west side of MF is a steep slope covered with rock debris. The rock at the top of the slope is from a landslide eons ago.

TERRACE FORMATION

The mountains of the Idaho batholith at one time were near the same elevation as the Rocky Mountains. Both have been reduced over geologic time to about half their original elevation. Geologists estimate most of the erosion took place during repeated ice ages. Glaciers are very heavy and good at crushing mountains into rock debris. One cannot imagine the flooding that took place when a glacier melted. The volume of melt water would have been measured in cubic miles. The speed of the running water would have been near the speed of an avalanche. The carrying force, bed load, of the river would have included very large boulders. Water at 62.4 pounds per cubic foot can move mountains.

Finding terraces at Pistol Creek is not difficult. They are everywhere. The difficulty is in trying to place the terraces in order of their origin. Terraces are depositional structures formed where deep river water flowed. When the water levels in the river decreased, the terraces became surfaces that were eroded. The thing to remember about terraces and their age is that the higher the elevation of the terrace the older the terrace. A terrace represents the bottom of the river. If you observe a terrace and it is fifty feet higher than where you are standing, the bottom of the river was once fifty feet above your head.

Once a terrace is abandoned by the river, it becomes an obstacle that must be eroded by a tributary that is now dammed in an old creek bed. The MF has eroded the present riverbed in the terrace on which the Ranch is located. The present tributary streams must dissect terraces so that the stream can do its job and drain its watershed. Try connecting the terraces that you observe at the Ranch and hypothesize about which terrace joined which terrace and how they came to be separated.

Cow Creek enters the MF across the river near the south property gate. Locate the Cow Creek terrace to the south of Cow Creek. It is flat, has burnt trees standing, and will be just above eye level with the Ranch terrace. Look for remnants of the terrace on each side of Cow Creek and join them. The north side terrace is not as obvious as the south. Notice how Cow Creek has eroded through the terrace almost in the middle. How could this have happened?

COW CREEK TERRACES, 2013

COW CREEK TERRACES

Cow Creek Mountain and valley is across the MF from the Ranch's up river gate. There are five terraces in Cow Creek valley where water eroded the creek bed through five terraces to empty into the MF. Notice the elevation of the highest terrace. Estimate its elevation above where you are standing. The top terrace was the river bottom thousands of years ago.

There is an anomaly on Cow Creek Mountain across from where Garden Creek flows into the MF. A geological anomaly is an oddity or mystery that is not fully understood. When the Pistol Creek Ranch "Firestorm" killed hundreds of trees on Cow Creek Mountain, boles (tree trunks) eventually fell and were strewn about the mountainside. Over the past twelve years the boles have settled to the bottom of the slope and are arranged in unique patterns. The boles weigh nearly a ton each. How have they managed to migrate to the bottom of the slope and be arranged in the patterns that we see them in today? If they had rolled they would still be strewn about, hung up on stumps and rock. If they had slid they would be pointing up slope. What is your conjecture for a solution to this anomaly?

BUCK LAKE (From the trail), 1990   PISTOL CREEK VALLEY
('V' shaped fire-scorched valley formed by water-flow)

PISTOL CREEK RANCH TERRACE

The PCR terrace is typical of terraces built by rivers in North America. The geology is simple. A terrace is a former riverbed, and its level surface is eroded rock and soil debris that were deposited in a basin or valley.

The present water flow is nothing compared to what it was thousands of years ago. The PC valley is filled with rocks rounded and smoothed by rushing waters of the older river. It is difficult to imagine the power of the rushing waters provided by glacial melt. Geologists estimate the mountains of the Idaho batholith, like the Rocky Mountains, were at one time twice as high. The rounded and smooth rocks found covering the terrace are proof that the MF once flowed where PR is today. The roots of the fallen, up-rooted trees around the ranch are clinging to rounded rocks.

Artillery Mountain's east slope where it meets with the MF valley reveals the mountain has undergone many huge rock slides as the MF eroded the river banks over thousands of years. As the river eroded at the base of the mountain, the mountainsides weakened & sloughed-off in large sections, like a tall stack of pancakes falling off a plate. Notice the terrace in the foreground. This terrace is located between Artillery Mountain & PC rapids. It was formed by the MF.

ARTILLERY LAKE (From the ridge and at Lake level), 1992

BOULDERS, PEBBLES, SAND AND CLAY   RANCH TERRACE

Where did the rock come from that caused PC rapids? The MF has eroded through the rock debris that dammed the river, reestablishing a new river bed. Note the mass of small loose at the right of the east bank of the river. Notice the steepness of the riverbank. Could the rocks that formed the rapids come from the eastside of the river? You bet' cha.

This terrace is near the Ranch "swim beach" on the Middle Fork. It is also thought by archeologists that the nearby hollows in the ground suggest that Indians camped here.

MIDDLE FORK EAST BANK (At Pistol Rapids)   CHOCOLATE MIDDLE FORK

The rockslide that formed the CP rapids originated from the bedrock shown at the top of the hill on the eastside of the MF.

Grim evidence of the firestorm that brutalized Pistol Creek still decorates the hills with blackened remains. Residual effects occur from time to time to remind us of the catastrophe. With the destruction of the foliage that firmly held the soil bank, landslides will continue until the soil is once again firmly held by vegetation. When the rockslides end in the MF, the river becomes a chocolate vehicle carrying soil. The sooner the river clears, the better for marine life.

EXPLODED GRANITE

JASPER (MINERAL)

This gray granite boulder expanded so quickly from the intense heat of the 2000 PC firestorm that it split open. This is common with granite. Beware of granite surrounding a firepit; it can explode when exposed to intense heat.

MINERALS AT PISTOL CREEK RANCH

A mineral has a distinct chemical formula, for example, Quartz, (SiO2). "A MINERAL IS NOT A ROCK." Rocks are made of minerals. There are few minerals (jasper and feldspar are two) in the PC area. There have been staked claims to mines with valuable minerals along the MF, but we have not seen a valid report of the results. Jasper is a mineral formed of silicon dioxide (SiO2) from quartz. It gets its color from iron. It is found on the Garden Creek Trail. Look for it on the face of the Lodge fireplace.

FELDSPAR (MINERAL)

HORNFELS (ROCK)

Feldspar is a mineral. There are two major feldspar minerals; gray plagioclase is rich in sodium and calcium; orthoclase feldspar is rich in potassium. Orthoclase feldspar is found in pink/red granite. Both feldspars show good cleavage: two faces will meet at ninety degrees, two sides will meet at less than ninety degrees. Feldspar plagioclase faces will meet at ninety degrees and two sides will meet at less than ninety degrees as shown above

Hornfels is a contact metamorphic rock. Ross Geiling, the hermit who had a cabin on 45 Creek, found hornfels on upper PC, and they are piled up at the burned-out site of his former cabin. It also gives a unique appearance to the Lodge fireplace.

ROCKS AT PISTOL CREEK RANCH

The rocks at PCR are igneous and metamorphic. Igneous rock is formed by the combination of two or more minerals. Keep in mind that rock is formed from molten magma that contains an abundance of minerals in different concentrations. When the materials cool, rock is formed from whatever minerals happen to be in the vicinity. This causes many named rocks to be closely related by mixture. However, an extra amount of a mineral in a rock can change its color and appearance, causing it to have a different name, for example, pink granite and grey granite.

The opposite is also true. One rock may be dark, nearly black and another may be light gray and have the same name, for example, basalt.

Two categories help us determine a rock's name, coarse grain and fine grain. When the crystals of a rock can be seen with the naked eye, the grain is coarse. If not, the rock is fine grain. Granite is a coarse grain rock. Fine-grained granite is rhyolite. Basalt is a fine-grained rock. There is not much basalt at PCR, as basalt is the rock of oceans. Granite is the rock of continents.

PINK GRANITE (ROCK)

DIORITE (ROCK)

Pink granite is igneous and intrusive. It contains quartz (clear/white), pink/orange orthoclase feldspar, and biotite mica in dark, brown sheets. Gray granite is more common in the Idaho batholith. It looks the same but is whiter from the presence of plagioclase calcium feldspar.

Diorite is a gray to dark gray igneous and intrusive rock composed of plagioclase feldspar that is white and hornblend that is black. The rock has low quartz content but contains several other minerals. It can also be blue-gray or black and is extremely hard.

DIORITE PORPHYRY (ROCK)

PATIO FLAGSTONE (ROCK)

Diorite porphyry is an igneous/metamorphic rock. It has metamorphosed plagioclase crystals. Porphyry refers to its tendency to be purple, but it can also have greenish or reddish coloration. The rock is volcanic in origin and was prized by Ancient Egyptians.

The stone is regional and metamorphic. It is formed from the deposition of volcanic ash, containing iron minerals, in fresh water lakes. The orange coloration is from the oxidation of iron minerals. The U.S. Forest Service flew this brittle and soft stone to the Ranch as a friendly gesture for rebuilding cabins after the fire.

ELK CARBONATED HYDROXYAPATITE

ROCK SLIDE   PLUTONIC ROCK

These rocks are igneous. This scree is near the south gate of PR. The jagged edges of the rock indicate it is from a rockslide rather than being transported by running water. The rocks were once above on the hill.

This rock is igneous and metamorphic. It is from the Atlanta Lobe of the batholith. It cooled, cracked, and was intruded by molten magma. The intrusion metamorphosed the original rock. A minor fault line is seen on the left. The Salmon River's Middle Fork did not deposit this rock at the horse corral.

$Ca_{10}(PO_4)_6(OH)_2$ Be careful when you explore Pistol Creek's geological features. You don't want to add your carbonated hydroxyapatite to the Ranch's mineral composition.

"If it is lying in a bed of rocks, if it looks like a rock, if it hefts like a rock, if it is marked like a rock, is it a rock?" Perhaps not, it could be a "nodule." A nodule is not a rock even though it has similar characteristics.

A nodule is a rounded mass or lump of an aggregate mineral without internal structure, a concretionary lump of manganese and other metals with detritus that was formed on the floor of the ocean. It is a chemical, sedimentary nodule, not a rock. (R.L. Bates & J.A. Jackson; Dictionary of Geological Terms, New York, Doubleday, 3rd ed. 1984).

In July 2015, I found a manganese nodule near the Pistol Creek Bridge and the Pistol Creek campground. The nodule was lying in a bank of rock debris near the top of the trail downriver from the bridge. It looked out-of-place in the surrounding rock debris because of its ochre color. Surrounded by rounded river gravel of igneous origin, it "stood out."

Physically the nodule was circular, about ten inches in diameter with excellent symmetry. When studied from the side it tapered from the center toward all sides, similar to a flying saucer. It weighed about eight pounds, the center was 2.5 to 3 inches in diameter, and it was circular, lenticular in shape.

What is the significance of this find? Relating back to the definition of a nodule, "they are formed in numerous oceans of the world." This nodule was found in Idaho, in the glaciated, deposited river-rock gravels of Pistol Creek, at an elevation of nearly 5000 feet above sea level. If the ocean where it originated was 500 feet deep, this nodule was found deposited 5500 feet above its origin. That is one mile higher than where it was formed. See page 63 for photograph of nodule found near Pistol Creel bridge.

Think about that! Where did this nodule come from and how did it get to where it was found? There are two scientific hypotheses for this anomaly.

1. Nearly 140 million years ago, near the beginning of the Cretaceous Geologic Period, the Niobrara Ocean, an intercontinental sea existing north and south, separated the entire continent and bisected the western North American Continent. This ocean filled the area of the continent that today is a mountainous region known as the Idaho Batholith and the Great Basin, and it extended into the Great Plains. This ocean existed for an estimated 70 million years and dried up near the end of the Cretaceous Period. Is it possible that the nodule I found at Pistol Creek originated at the bottom of the Niobrara intercontinental ocean and was lifted to the high elevation when the batholith was raised and formed by internal forces within the earth's core and mantel? This is a long-shot hypothesis, but possible. If so, this nodule is at least 110 million years old, give or take 20 million years. The Idaho Batholith was formed during the late Cretaceous Period, so the nodule is many millions of years older than the rock debris of the batholith in which it was found.
2. The nodule was formed more recently in geologic time after the Idaho Batholith was formed. The womb for formation could possibly be glacier melt water in a fresh water lake or stream in the mountain region. I have not found information supporting fresh water formation for manganese nodules; it is only known to be oceanic in origin. Manganese in superheated solution is released onto ocean bottoms from volcanic vents, which causes the formation of nodules. Well, there is another remote possibility for the nodule. It was "salted" by a practical joker, who will get a laugh out of it being found and identified by an earth scientist.

HOT AND COLD SPRINGS NEAR PISTOL CREEK RANCH

We have noted the many geological features such as rocks, rockslides, soil slumps, faults, muddy river water, rounded mountains, and V-shaped valleys around PR that catch your eye. It is also worth noting hot and fresh water springs with surrounding plant and animal life.

NODULE FRAGMENT (11 CM), 2015

HOT SPRINGS

The hot springs in Pistol Creek, above Little Pistol Creek, are worth visiting. Elk hunters know they are a favorite elk wallow, but they are also an interesting geological feature. Hot springs are produced by water heated by high temperature rock material situated superficially under the creek bed. To produce steam, water in the aquifer, replenished by the surface flow, seeps down through fissures in the bedrock to the geothermal heated hot rock area where the water is heated to boiling and forced to the surface under pressure. The super-heated water will remain liquid until the pressure is reduced near the surface. At the surface the super-heated water is released as steam. This is the same process that produces hot springs and geysers in Yellowstone National Park only on a much, much, smaller scale.

COLD SPRINGS

Fresh water springs are abundant near the Ranch. One located upriver, outside the Ranch gate on the Middle Fork trail, is a prime location to see wild orchids. Another cold spring on Garden Creek is an excellent source of cool drinking water. How do these cold-water springs develop? Water in the higher mountain regions infiltrates bedrock fissures and flows underground to cool ground areas at lower elevation outlets.

These springs provide many of the wet lands that support abundant wildflower growth. Where the bedrock provides a sub-surface basin, marshes are formed. During warmer weather an upper elevation stream may empty into a marsh, producing continuous water flow to the Middle Fork.

Rocks are Beautiful!
Rocks are also for Collecting!
Help A Youngster Make A Beautiful
Rock Collection, But Remember,
Rocks Are Also For Throwing,
So, Be Sure To Duck!
Rock On, Kids!

REMINISCENCES OF PISTOL CREEK RANCH

Billie and Clint Gerlach: "When Marv Hornback purchased the Ranch, his plan was to build cabins and sell them to obtain money to build the next one. He initially built a triplex lodge with the intention of renting rooms to hunters and salmon fisherman. In those days, you could catch and keep fish you caught, and rules were far less stringent than today.

'Marv, wife Barbara, along with Dewey Heater and son, Don, built the first cabins. The Ranch was not without its problems. The first home of the Hornbacks was destroyed by fire. It was located where the Thornton's cabin is now located by the MF.

'The tree house in the pasture by our cabin was built for the Hornback girls. Patty, the eldest daughter, was a real cowgirl, and there wasn't much she didn't try-floating the river in an inner tube or riding her motorbike at full speed.

'Marv went up in a homebuilt airplane with a friend at Strawberry Glen Airport near Boise and died in a crash. As a result, the Ranch was in peril of being lost by those who had purchased cabins. This was before we bought our place in 1967.

'Those who owned cabins decided to buy the Ranch from Barbara and make it into an association. Bruce McKeighan was instrumental in that. He lives in Escondido, California.

'Clint is a hunter, but he has emphysema and can no longer hunt or go into the cabin because of the altitude, much to our sorrow. We love the Ranch and look affectionately on the many years we spent there.

'The big fire in 2000 destroyed many of the animal mounts he accumulated over the years in Africa, Alaska, Canada, and at the Ranch. We keep the Ranch for our grandson, Nicky, son Curt, and wife Roberta.

'Our introduction to the Ranch was by John Chapman, who was a friend of Bruce McKeighan, who first invited us for a hunt. The Ranch and its way of life intrigued us. I am not great on sleeping on the ground but do love the mountains. It was perfect for us. Clint hunted and I would read, ride horseback, and hike.

'Nancy McKeighan and I spent our summers walking ten miles a day. We went downriver five miles and back again. It was great fun.

'Most people leaned about the Ranch by word of mouth, for the most part, in the old days.

'We bought our furnished cabin in 1967 for $18,000 and paid another $3,000 for the common stock. When we were looking to buy, we looked at the cabin we now own and C10, which was for sale for $12,000 by its owner. Cabin (C10) was bought by Brace in 1969.

'A few years after purchasing our place we added a game room and enlarged one bedrooms. I don't recall the add- on cost. Our cabin in 1967 was the last one Marv built, so it was the newest one at the Ranch.

'Al Grant and his wife, Donna, were friends of ours. They lived in Pacific Palisades, California. They were our guests for several years before they bought their cabin. Donna was a great cook, and I don't think anyone could have loved the Ranch more than Al. He died of cancer shortly after his last visit there. Clay Lacy took him for a tour of the woods in his helicopter. I can still see the peaceful look on Al's face as he flew over the woods he so dearly loved. Donna sold the cabin after Al's death and she, too, is gone.

'Clay Woods was the Ranch manager for several years. He, wife Marsha, and daughter enjoyed the Ranch. When he died either his wife or daughter contacted Scott Patrick and asked that Clay's ashes be flown to the Ranch and scattered there. I was not there when this happened, so I am only repeating what someone told me."

Meliss (Brown) Clark: "Where the lodge is now located, I think there was a cabin built by Dewey and Don Heater. They were the head carpenters. I think that cabin burned, but I don't know when.

'Where the dump is now located, Dr. Jim Scott from Redding, California, built a big cabin. During hunting season, while they were at the lodge eating dinner after a day of hunting, they hung long johns or some other type of clothing along the mantel to dry. After dinner, they went back to a cabin in ashes.

'C7: Dean Brown, head baker for Albertson's in Boise, would fly in Friday, get a pot of his "famous stew" going and have a few libations, with or without company.

'The early days were like the Wild West up at the Ranch. That's not to say it was right or good, but that was how many operated, which essentially meant borrow, break, and maybe, or usually, not replace. Invasion, use of unoccupied cabins, and breaking game laws was the rule. I know there are a few who would like it to be more like the OLD DAYS, but no one took care of equipment and so forth- not really a good plan. It is better now when most follow reasonable rules, etc.

'I was absent from Pistol Creek from the spring of 1965 until the summer of 1982. I trailed horses in from Salmon with Bill Watson, my husband (no relation to Eleck Watson). Bill was also a sawyer. He logged using a wagon and horses.

'For six weeks in the fall of 1963, I stayed in cabin C22 with three little ones. It was the only occupied cabin on the ranch. I did my laundry in the bathtub. I would read late into the night in the bathroom with packrats running around beneath the tub. It was the only time I asked to leave the Ranch and be taken to town. I was a little overwhelmed.

'As to Floyd Posing-I know he worked with Bill Watson sometimes. He worked on the river, and I think he guided at least one fall when I was in the ranch.

'I was in the ranch the fall-winters of '60, '61, '62 (that's when I rode out with Bill and the horses downriver— VERY COLD, BELOW ZERO), '63, and part of '64 (two to three weeks into hunting season).

'The cook had to be taken out to town because he was a drunk. I was pregnant and could barely stand the smell of pancakes, so Bill helped with breakfast. Other than that, I cooked."

David Pecora: C1: "Trivia note regarding the A-frame: Buzz Chaney built the cabin himself, virtually unaided. He was also one of the best sprint car, motorcycle, and stock car racers in the western states and had the broken bones to prove it. I remember him tearing down the road behind our cabin on his big dirt bike at blinding speed. The man lived on adrenaline. Buzz was the Ranch manager in the early '70s. Ferris Lind sold the lot to Chaney who built the A-frame.

'C2: Dick DeLong originally owned the red, white, and blue jeep. Evidently it was "used by the ranch." I remember my dad tell of how they cut the jeep body and frame rail into four pieces to fit in the Cessna 180, Mar welded it back together. Robert Boyd bought the jeep, and had it flown out. I think he still owns it.

'C3: Purchased from Hornback July 12, 1962. The lot cost $3,000, cabin $9,500, and Pecora bought Smilanich out in 1998.

'Dennis Smilanich flew in the China/Burma/India during WWI, flying the "hump" across the Himalayas and lived to tell about it. After the war, he was one of the first pilots for Empire Airlines. Empire Airlines grew out of Zimmerly Airlines of Coeur d' Alene, Idaho, in March 1946. Empire was still flying three Boeing 247s until March 1948 when they upgraded to Douglas DC-3s. They flew mainly in Idaho serving Boise, Lewiston, Pocatello, and Coeur d'Alene. Routes were later expanded to Washington and Oregon.

'In the mid-1950s Empire was absorbed into West Coast Airlines. West Coast, which hired my dad in 1955) eventually merged with Bonanza and Pacific Airlines to become Air West, which was bought by

Howard Hughes and became Hughes Air West. Dennis retired from Hughes. I was at his retirement ceremony in Seattle. His copilot seemed glad to see him retire. Hughes eventually merged with Republic Airlines, which was bought by Northwest, and was later bought by Delta. Dennis also owned several successful businesses. He owned a Beechcraft dealership, a charter operation in Boise, an Avis car franchise, and an aerial firefighting operation, which used war surplus bombers. One of his B-25s was used in the film, Catch-22.

'C4: Capt. Turner was a Pan Am Clipper flying boat pilot in the '30s and early '40s. He was in command of the flight from SFO to Honolulu on 7 December 1939 and was scheduled to land about the time of the Japanese attack on Pearl Harbor. He was diverted to Maui per secret orders. After refueling the plane, he loaded critical personnel and left about dark to scurry back to the mainland. He flew Clippers and various land planes for the US Army Air Corps during the remainder of the war.

'C6: In 1980, Grammar and Swanson purchased lots 38 and 42 (where C6 is built) from Pecora / Smilanich, who had previously purchased the lots from Adelaide Marshall (Wayne) in 1972.

'C7: Dean Brown was a first-class great guy with a heart the size of Big Baldy.

'C14: George's last name was Dovel. Don Boyd of Portland, Oregon, bought his cabin in 1969.

'C16: Ferris Lind was stricken by polio and became totally paralyzed. He visited the ranch in an iron lung. Lind also owned lot #1 and sold it to Chaney.

'I also have various documents relating to the sale of the ranch to the MFR owners after Marv's death. Apparently, a significant portion of the ranch was still owned and controlled by Adelaide Wayne (Marshall) when Barbara decided to sell. I believe this stems from the fact that the Hornbacks were still making payments to Wayne.

'Other court documents are revealing. Some excerpts: "Pistol Creek Ranch is a homestead.... A portion was subdivided into thirty lots, of which twenty-two have been sold...."The Hornbacks" are purchasing the ranch under contract of sale dated Sept 8, 1958, from Mrs. Adelaide Anderson Wayne (a widow)...." Wayne has "conveyed title to twenty-two lots, together with easements to land upon the airfield...." It was agreed that "no more than 30 commercial cabins were to be built upon the ranch."

'My dad's warranty deed (the filing fee was 90 cents!) names Mar and Barbara as grantors, and there's a separate easement with Wayne for access to the airstrip etc.

'In August 1966, about the time MFR Inc. was formed, the owners signed an option with Barbara in which they held exclusive rights to purchase her equity in the ranch for $70,000. The agreement mentions that Barbara held a "vendee's interest in the premises under contract from Wayne, and that the purchasers could assume this interest in lieu of title free and clear. In other words, the owners were purchasing Barbara's equity and assuming her contract with Wayne. I have a 1966 document from McKeighan (our first president) in which he declares the formation of the corporation and mentions the immediate need to raise funds, $20,000 for Barbara Hornback, $5,000 for Wayne, and $5,000 for ranch operations. The April 1968 balance sheet shows a short-term mortgage of $10,000 due in one year, and a long-term mortgage of $30,000. By 1969, the long-term note was down to $20,000 and there still remained a short term $10,000 obligation. MFR was taking out short-term loans and making lump-sum payments to either Barbara or Wayne. An interesting note in a Dec. 1969 letter to owners: "We have paid Mrs. Marshall $11,000 mortgage/interest payment, leaving a balance of $10,000 due end 1970, closing out our mortgage obligation on the ranch.

'Wayne evidently remarried, became Adelaide Marshall, and would seem to have three last names: Anderson Wayne Marshall. I don't have enough info to confidently backtrack the finances, but it would seem that Barbara was paid off either immediately or very shortly after MFR. Inc. was formed. Marshall (ne' Wayne) was paid off in 1970."

Harold Dougal (by APS): "Harold Dougal logged more than 20,000 hours flying between his first flight in 1937 and his retirement in 1989. His career spanned from a twenty-year contract to patrol power lines and pipelines, to 1956- 57 when he was chief pilot for Aircraft Service Company of San Francisco, to the early 1960s until 1981 when he worked for Boise Air Service, and to 1989 when he retired after a three-year stint as chief pilot for Scott Patrick Aircraft. His book, Adventures of an Idaho Mountain Pilot, is available from Harold Dougal, 1115 Shaw Drive, Boise, Idaho 83705, and we highly recommend it for its history of Pistol Creek Ranch, flying the Middle Fork, and life along the river.

'Harold Dougal also helped his friends, Bill Wayne, and later, Marv Hornback, build the Pistol Creek Ranch. Before Hornback built the runway at the Ranch, Dougal flew supplies to Indian Creek and drove them (before its wilderness designation) with a tractor and wagon to Pistol Creek.

'In his book, he recalls original Ranch owners: Bill Pecora, Dick DeLong, and Clint Gerlach. Dougal gave the authors permission to use photographs and information from his book in Idaho Wildflowers at Pistol Creek (2013 edition)."

Pat Hawn, Ranch manager: "In 1997, after the little memorial service for Al Grant at the Ranch, Clay Lacy took his ashes up in his plane to scatter them over the runway. Well, when he opened the window to dump the ashes out, most of them blew back into the plane. After he landed and everybody went back to their cabins, Clay asked for a vacuum cleaner and proceeded to vacuum Al's ashes out of the plane. I took the ashes and placed them at the side of the runway."

John Schumacher: "Regarding cabin C21, my father, Jack Schumacher, and his insurance company partners, Daryl Turner and Albert Erkel, bought our cabin in 1966 from three Mormon men from Southern California who all went through a divorce at the same time and had to sell because their wives took all their money.

'A funny story, my father was told by Daryl Turner that the cabin was for sale and if they were interested, they needed to fly up in the next five days with a cashier's check deposit. My father cancelled a scheduled trip to Hawaii with my mom for their anniversary three days before they were to leave to go to the ranch and see for himself and present the deposit...and catch some steelhead salmon and drink a fair amount of whiskey from what he told me. My mother was not happy but thank God, he did that.

'The ringleader of the three original owners was a man named Richard Russell who was an accomplished California state senator from Pasadena. They were all members of the John Birch Society and there was eighteen months' worth of dried food stored in our attic in case Russia came calling and the cabin owners needed a place to hide.

'Don Heater and his father originally built our cabin in 1960. The insulation used in the walls was sawdust. There were three floor plans to choose from and the largest four-bedroom plan was the one that was build. My father and his partners bought cabin 21 in 1966 for $20,000 a large sum in those days, and I have the bill of sale."

Scott Schumacher: "My brother, John, forwarded me your email below about Richardson, the guy who was an original part owner of our cabin, C21. On page 11 is a copy of his watercolor of the barn, etc.

'The original hung in our cabin until the fire of 2000 when it burned down. Luckily, I took this photo before the fire, and I have it framed in my den at home. Maybe we can track down his family and get them to buy Paulson's lot; I'd at least like to forward them this image so they can have it."

Robert L. "Rusty" Brace: "William A. Brace flew to Indian Creek in John Conroy's DC-3 in October 1969. (Conroy owned cabin C18 with Dr. Jim Scott.) Cabin C10 was for sale by the Ross Brothers, Leonard and Raymond, who signed their deed in Plumas County, California. Upon returning to Santa Barbara, Bill searched for partners to invest in Cabin C10."

Meredith Brace: John Lancaster Accident: "I was sixth months pregnant with my second child, so I remember being somewhat more uncomfortable on that hot day. I was in the front room of Cabin C10 and was facing the river talking to my friend, Jane Fredericks, who was on a barstool facing the runway. We had heard the plane take off, and the next thing I remember was seeing a shadow of a plane behind Jane float over the river, against the steep hill. I also remember not hearing an engine, just seeing the shadow. We mentioned to each other how strange that was, but kept our regular conversation going. The next thing we heard was a huge crash and looked toward the runway and saw billowing smoke. Jane is an RN, and her instinct was to grab her first aid kit and run over there while I stayed with her daughter (3 years old at the time). Of course, by the time she got there, there was nothing she could do."

Patty (Hornback) Vance: "C2: We called it "The DeLong cabin." Dick and Betty DeLong, Chet Lancaster and a third owner named Smith. Can't remember his first name.

C5: Clint & Billie Gerlach bought the cabin we called the Reno cabin. First owners were Marv and Gary Whiteman from Nevada. Marv still lives in McCall, ID.

C7: Dean Brown & I think Joe Albertson was part owner. He came to the ranch with Dean several times.

C9: Dr. (Charles) Fleming and wife in the beginning was Shirley, children Linda, Jan, and Chip.

C11: Gene Barton who later bought Sulphur Creek Ranch.

C12: Bruce McKeighan, Ray Nixon, and Duane Tjumslond were original owners.

C17: Estelle and Dale Dooley and later sold to Dr. Earl Scott.

C19: Allen Paulson, Clay Lacy, and Dave Robbins were the original owners.

C20: Jack Mattich from Calif. was the original owner. Clay Lacy purchased it later.

C21: The last cabin was called the Coleman cabin and one of the owners was a politician from California named Richardson. I only remembered because he painted watercolors and did one of the barns and runway. He threw it away and I dug it out of the trash (Thank God), and I still have it framed in my den.

'We (Jan Fleming and me) hung out as kids at PCR. I was the leader and always the troublemaker. It would be great to talk to her again. Sometimes our stories grow more exciting with the passing of years, but you might ask her if she remembers us (Jan, her sister Linda, and me) burying vodka and selling it to the smoke jumpers at Indian Creek. We were real entrepreneurs; we got $17 a bottle in the '60s. Her mom was a very nice person and always nice to the kids."

Bill Widgren, Ranch manager: "I started at the Ranch in fall 1990, but I missed the hunting season of 1991. I came as a steady ranch hand in spring 1992. Sonja and I started as Ranch managers in spring '97 and stayed until late fall 2002.

'John Lancaster's crash. He left the ranch down river, turned 180 degrees, and climbed out about even with the Lodge heading up Pistol Creek when we heard a bang and saw smoke coming out the exhaust.

'We happened to be standing by the fuel shed watching him climb out when he immediately turned back into the landing pattern, but because he wasn't high enough, he couldn't get far enough away from the runway as he turned on base to line up with the runway. His engine was sputtering all the way around but conked out on base turn and he couldn't glide. He was between the runway and Grammars. Those on the ranch / cabins were A. Sohn and guests (5), Brace (7), and Casey (2), plus a couple of ranch crew."

Bruce McKeighan: February 14, 2013, telephone interview by APS: "Joe Albertson bought Cabin C7 and gave it to Dean Brown.

'Ray Nixon is my wife's cousin, and he lives in Downey, California. Duane Tjumslond was a cement contractor in southern California. They were partners in my cabin C12.

'After Marv Hornback died in 1965, a lady in Boise had a note for $70,000 from the Hornbacks for the Ranch.

'I put together the deal for each cabin/lot owner to pay $3,000 a piece to pay off the note and own the Ranch. Some owners were reluctant to buy into the Ranch because they didn't understand why it was necessary.

'In 1980, I built the log cabin C6 that Grammar bought. He paid around $100,000 for the cabin and lot.

'In 1980, the US Forest Service had an interest in the Ranch, but they made no offer. They paid the Ranch not to build on vacant lots on the Middle Fork riverbank."

Lori (McKeighan) McKenna: March 12, 2013, telephone interview) by APS: "I am Bruce McKeighan's daughter, Lori McKenna. I was visiting my dad last weekend and he told me of your quest to find information on Pistol Creek. He asked if I would be able to help you out with stories and photos since he is now blind with macular degeneration and can barely hear, making it difficult to converse with him over the phone as you may have discovered. I have innumerable photographs and the original blue prints from when my dad acquired Pistol Creek. The original plans are missing information, which I may be able to fill in for you. We started going in to the ranch in summer of 1962 and did so every year for the next thirteen years. My dad was going in for hunting season even before that. The last time I went into the ranch was around 1977.

'John Chapman took my dad along to a hunting show at the Pan Pacific Hotel in Los Angeles. Marvin Hornback had a booth there promoting hunting in the backcountry. He promised if they came in they would bag an elk. They did! What Marv would do is fly his plane over the area until he found a herd of elk then pack the guests and his men and horses to that place for a hunt.

'My dad first bought our cabin for $13,000. He went in on it with Bradleys, Nixons and Tjumslonds. My dad's partners, Nixon and Tjumslond, only came to the Ranch to hunt.

'My dad was an excavation contractor in southern California.

'Julie and Don Peters were early Ranch Managers.

'My dad bought the Ranch, incorporated it and sold stock to lot/cabin owners, who became owners of the Ranch. "The lodge at the swimming hole was the focus of Ranch activities during the 1960s. It had rooms for rent, a commissary, laundry, and served food.

'In the pasture between the lodge and barn was a large tree house for kids. It had three bunk beds and was a big deal. We went there and smoked our first cigarettes.

'Horses, ponies, hogs, and chickens were flown to the Ranch.

'From 1962 to 1975, my mother, Natalie McKeighan, my sister, and I would spend summers at the Ranch.

'The highlight of the summer was the Fourth of July party. Clay Lacy would take us for acrobatic rides in his airplane. He would go straight up and straight down.

'Another remembered activity was telling ghost stories in the cemetery behind the Hornback cabin. The Hornback cabin burned and cabin C22 now occupies the site. The Ranch cemetery is located behind C22.

'We used to fly in my Dad's Bonanza, from Van Nuys Airport in Los Angeles in the very early morning on our first day of summer vacation. We didn't return from the ranch until the day before school started again, just in time to buy our new knee socks and school supplies. My dad, Bruce McKeighan, my mom, Natalie, and my sister, Joni, and I would make the four-hour trip, usually refueling in Boise. Then 45 minutes on into the ranch. Often our other sisters and assorted friends and family would spend time with us at the Ranch as well. Most would make the 14-hour drive from Los Angeles to Boise, then Ray Arnold or someone from Boise Aviation would fly them in, or Marv Hornback or my dad would pick them up. I have photos that start around 1965 and go to 1976, which was my last year at the Ranch. I went in for the 4th of July celebration and took my first baby. My biggest regret is that I never had the chance to take my husband in, so he's only heard the stories and seen the pictures over the years. There are a few even older photos, going back to the days when my dad, Nixon, and Tjumslond went in mainly for fishing and hunting. In those days fish were plentiful and the river high. You were allowed to keep your fish. We always caught our limit of fish, and enjoyed hearty dinners of fish and potatoes. We buttered the fish up Irish style, sprinkled a bit of Lawry's salt on them, wrapped them in tin foil and threw them on a roaring campfire. In this way, we had some of the most succulent suppers to say nothing of the camaraderie of friends around the campfire. I still remember with glee, rolling with laughter from bawdy jokes the cowboys would tell and being lulled to sleep by sweet songs played on guitars country style until long after the sun went down. Usually the young rangers would come from the station at Indian creek to meet up with all the cute girls who were around for these forays. They played guitar and in later years I too learned to play. Namely, my older sisters, their friends and some of the regulars like, Patty Hornback were the main attraction for the rangers, and I was still just a kid. Patty Hornback was the most memorable of all! She, her sister Jackie, father Marvin and mother Barbara comprised a huge part of my wonderful childhood memories at Pistol Creek.

'On two occasions, we had prominent politicians visit the ranch and over the years a celebrity or two as well. The Robert F. Kennedy clan, including a slew of kids my own age came into the ranch. They spent the night at PC Hole where they proceeded with a float trip down the Middle Fork. A raucous good time was had by all as they included us in their first day and campfire that night. I specifically remember showing the kids all the best rocks to jump off of into the deepest swimming hole.

'Back then, we referred to Bill Watson as "Wild Bill" Watson. He was the main cowboy-hand at the ranch. He lived in one of the rooms they rented out at the Lodge. One of his helpers was a popular fellow named Tommy Chalupca. Unfortunately, Bill drowned when he had a heart attack in waist high water years later. Tommy is still in Boise and recently attended my mother Natalie's funeral along with Patty Hornback (Vance), who remains a dear friend to this day. I only mention them because when the Kennedy's came, they were awestruck by our "real cowboy" and "cowgirl" and it made us puff up with Pistol Creek pride!

'We had plenty of horses for rent up at the ranch back in those days. The grandest time could be had hanging over the rails of the corrals by the barn watching the cowboys break horses. It was a real Wild West show. As we clicked our boots together and tried not to fall off the fence, we watched with unbridled glee and fear all at once. Pistol Creek was always a place of extremes.

'Around 1968 we started bringing our own ponies into PC. Marv and my dad first brought "Bullet" in for Marv's daughter Jackie. Patty already had a large horse called "Roxy." Bullet was an extremely ornery Shetland Pony with a mind of his own. My sister, Joni, and I soon got our own ponies as well. They were Welsh Mountain ponies, which are just a head bigger than Shetlands but just as ornery. Joni's used to run to the nearest mud puddle and lie down trying to roll off whomever was riding him. "Spooky" also liked to run close enough to a stand of trees to brush you off with an unceremonious thump! We always put the new kids on him to see if they could pass our own toughness test. Many tried but most failed. We all had a lot of gut busting laughs! I spent hours every day riding my pony, "Lady Bird" (named for Lady Bird Johnson). While I owned her she foaled two more ponies, "Pistol" and "Popsicle". I usually rode bareback, but I still have my saddle in my possession today. I used to race the planes when they landed, right on the dirt path beside the runway. I'm guessing someone put an end to that now, but boy could I run that pony! I lounged in the pasture with her; sometimes I just lay on her back looking up at the clouds. She took me to al the best locations, including Froggy Pond where I was obsessed with catching frogs all day. And yes, I brought them home. I called it, "The frog relocation program." Those were blissful, idyllic days making my early childhood every child's dream.

'Pistol Creek brings back the best memories of my childhood. Hot dusty summers when the ground would singe bare feet. I can still feel it because it was bare feet or cowboy boots and nothing in between. I spent my summers riding my horse along the river paths from Pistol Creek up river to PC Hole and down to Garden Creek and Indian Creek. We built many a dam trying to make Garden Creek into a swimming hole! Did you know that Brown's Hole (just beneath the crow's nest at his cabin) had the best trout fishing? It was hard to get down, so it never got fished out. Those of us who loved to fish knew it though, and I was one of them. I even loved to clean fish and often offered to do the duty for another fisherman--for a price!

'One summer I decided to gather kids from all the cabins and put on a play. I always had something up my sleeve. We put on the greatest production of "The Sound of Music" ever played out in Idaho's backcountry. For several summers, I became a teacher and in a quiet, forced march, gathered kids from near and far to come do things I thought would be fun, like, workbooks, but always with the promise of an art project at the end. I wonder now how they all really felt about their 12-year-old teacher!

'Ah well, it was something to do that the parents approved. I certainly had no lack of participants at the time and maybe that's why I later became a teacher, a mother, a foster mom and an adoptive mom. Ten kids in all! Perhaps my love of teaching and making my own art and homeschooling my kids all started because of summers at PC.

'It wasn't all work though. Most of us kids, I guess I should only speak for myself here, had our first kiss, our first (and last) cigarette and got drunk at least once up at our magical place in the woods. Without electricity, Facebook, video games and such, which kids today tune in to, we were left to our own devices and took full advantage of the freedom we incurred at our favorite place in the whole world, Pistol Creek on the Middle Fork of the Salmon River.

'The Hornback's old cabin at the beginning of the runway and across from the barn and airplane hangar is where the graveyard was located. We used to sneak out late at night with flashlights and meet there to tell terrifying ghost stories handed down through many summers, and scare the pants off each other. Many of our antics took place in the tree house, that Marv Hornback built for his girls. It was between the old lodge and the barn, right on the creek. There was a wooden staircase leading right up into a large one-room fort.

'There were three bunks, and we were allowed to spend many a night up there. We used to lead Bullet, Jackie's pony right up those steps and he would sleep standing up right there with us in the tree house. That too, for the pony was a forced march. I think we were just as ornery as the Shetland. I wonder if there are still the great and mighty thunderstorms up there that we used to encounter. One could be off lazing their day away a mile or so downriver and suddenly the sky would turn dark and ominous, and you knew with certainty that downpour would come before you could gallop home. You'd return to your cabin looking like a drowned rat. That meant more laundry due to a second change of clothes for the day, which I imagine now didn't make our mothers too happy.

'Laundry day now those are some great memories for me. We thought it was all great fun and a dandy treat even if our moms would beg to differ. If you didn't do your laundry at your cabin by hand on a washboard like they sometimes did, you took it down the lodge where you could pay them to do it for you, or you could use their laundry house. The laundry house sat back of the lodge by the river and the washer was a machine you got hot water to by lighting a big fire in a stove and warming it up. It would go through the machine, fueled by propane and then we'd take it out, put it through a hand turned wringer and hang it on one of the five lines they had out. Remember, this wasn't the 1800s; it was Pistol Creek in the 1960s. Laundry day was such fun that I used to go help Barb, Patty, and Jackie on laundry day. One could get their laundry done for a price down at the lodge. Our reward was an icy cold glass of red Kool Aid, which I can remember vividly as being the sweetest treat of all time.

'There were few private pilots around when we spent our summers at PC. Most families flew or drove to Boise and came in on charters. My dad, Marv, and Clay Lacy always flew us in by themselves. My dad was my hero because he was one of only two pilots (the other being Clay Lacy) who would buzz the runway. It was quite a thrill and made you feel invincible. Sadly, whilst we were enjoying our thrill rides and keeping safe, many didn't savor the same fate. There were many summers that someone would come or go, usually in the heat of the day and not familiar with flying the backcountry.

'They ended up nose down in the river or implanted in the side of the mountains. The forest service always retrieved the bodies and held them in the hangar at the end of the runway to be flown out next day.

'While their fates were tragic and sad at the same time, they fueled the fire for irreverent kids to tell even more horrifying ghost stories. I can't watch scary movies to this day, having had my share of reality back then.

'There was a lot of partying going on back in those days. Now it's the parents I'm talking about, not the kids! Cocktail hour started at 5 pm. Our parents fondly called it, "Having a little popsicle." More hilarity evolved in some cases, and some things transpired that I don't like to talk about to this day. The chaos of adults could be terrifying to a kid. One of the stories in between hilarity and terrifying was when there was a hardy, party in full swing at our cabin. Things were getting out of hand when Wild Bill rode his horse right into the kitchen and reared him up onto the kitchen island! I remember Mike Paulson, whose dad Allan Paulson owned a cabin as well, and I were hiding under a bed in the bunkroom timidly watching the whole scene. We never saw alcohol infused partying in the same way after that and to this day I only have a glass of wine with dinner about twice a year.

'As for Mike Paulson, he and I have remained close friends for the past 50 years. Once we spent an entire summer trying to build an airplane from his brother's old motor parts. Mike and I lorded it over the other kids as to who we were going to let fly in our airplane. We of course were going to fly to Pistol Creek! We went up and very far away but only in our own minds.

'Mike's eldest brother Bobby was married to Patty Hornback when they were about eighteen. Bob later died in a plane crash and after attending the funeral, my dad flew Patty and her daughter Crystal to LA to live with us for about 3 years. Crystal and her husband Garth were bequeathed their cabin by Allan Paulson upon his death. They are still enjoying it today along with their daughter Savannah, after rebuilding it after the big 2000 firestorm.

'Our cabins were lit with propane lanterns attached to the walls in each room of our cabins. We lite them with a wick, which transferred into a soft, yellow glow for light. A small propane heater lit the stove and kept the hot water and fridge running. It was a long wait for the dads to fly in with propane if we ran out in the middle of the week. And remember, no cell phones or computers for communication. If it was a real emergency, one would walk or ride a horse up to Indian Creek to use the Rangers' phone. In the case of running out of propane, I remember cold showers and then standing in front of a blazing fireplace trying to get warm. Nights were still cold, even after hot days up at the ranch.

'When my dad would fly in on weekends, we would run to watch him buzz the runway and then pull our little red wagon over to the spot on the edge of the runway closest to our cabin, to help unload the cargo. It was like Christmas every time! Jackie had a pony cart, which we also hooked Bullet up to and helped out when anyone landed and needed supplies carted to their cabin. We were nothing if not helpful kids!

'My dad helped build the old sawmill and a couple of the cabins in the later days. He was very handy and could build sheds and picnic tables and add porches on to anything. He always had a project going on. Our porch was one of the first that he extended down several steps to overlook the river.

'As you can tell by now I was a wee bit of a wild child. It had to be the Irish in me, along with freedom most kids only dream of. I instigated the big squirrel catching enterprise. We used to put a box propped up by a stick in the woods. The stick was attached to a string that led to our hiding place behind the nearest tree. The box was littered with peanuts, which were laid out in a trail leading to the trap. One could sit for hours, perhaps until dark waiting patiently for her catch. Once we caught one, we would reach in and grab it, usually getting bit or having it escape. Some of us even got ours into cages we built of wood and chicken wire.

'We fed them well and took good care of them all summer, letting them free before we flew home. It was really quite humane actually. I can't believe that none of us ever got rabies.

'I got my first dog at the ranch. I was such an animal lover and some of the hands had dogs that herded horses, but there were always a few strays there too. I took to one and had to make her mine. I was too shy to ask, so my mom and dad did so one night when they went to a party down at the lodge. When I found out that dog was to be my own, my life was never the same. I loved the heck out of her until the day she died 14 years later. She was just a little mutt, but a terrific one. The cowhands had real sheep dogs and put on quite a show when they herded horses. Those dogs were amazing. They remind me of the sheepdogs today, in my beloved Ireland, where I go to retreat on the Berra Peninsula, where they are still working dogs to this day. Our cowboys loved those dogs just the way the shepherds do in Ireland. The Berra is one of the only places I have ever found that reminds me of the wilds of Pistol Creek. No wonder I love it so.

'We spent a week every summer on a float trip down the Middle Fork. A cowboy would guide us, and it was always a thrill ride, just praying you didn't hit a huge rock sticking out in the middle of the river and get knocked over. I remember once my brother-in-law throwing me out of a raft and swimming for my life.

'All in good fun! We kids also used inner tubes to float between Pistol Creek Hole and Indian Creek, which was about a 4-mile trip. Man-Oh-Man, those were the days. Once, Kim Gerlach broke her foot and had to be flown out to get a cast. That was an exciting summer. Another amusement for us was the various pack trips we went on. We would start out in the early morning with our horses and a mule for packing. We went out for two or three days at a time. Once my saddle fell off in the middle of a very fast-moving creek we were crossing, and I almost drowned not being able to get to shore in a timely manner. It was my own fault for not cinching my saddle properly, and a good lesson was learned. Once in a while when we were camping, a tethered horse would get away, and it was always an adventure to catch him before he got back to the barn, which was the first place they always headed.

'In the later days, when our parents were divorced, we kids still wanted to go into the ranch. So, our parents hired various sitters to spend a month or so with us at our cabin. We often brought a friend or two, and Mike Paulson's parents let him come in with us. One of the sitters was Joella Watson, a niece of Wild Bill's. A family friend, Tony, was once in charge of us and she fell in love with the ranch so dearly that the next year she bought a horse there and named him Stony after a song called Stony End that Barbara Streisand made famous. Tony died an untimely death at the age of 26 from lung cancer and Ms. Streisand who was a friend of Tony's father's, eulogized her and sang Stony End at the funeral. There wasn't anybody who ever went to Pistol Creek who wasn't touched by it.

'At the end of each summer we sadly packed up, and with our shoulders hunched and our faces frowning, maybe even a tear or two as we took off, and flew back to LA to start school and real life again. I wrote a little song that all of us sang for many, many years whenever we had to leave. It went a little like this: "Off we go to Idaho to Pistol Creek and back again. We won't cry or shed a tear cause we'll be back again next year. Yee Haw So."

'There are many more stories to be told. I will regal you with the time my dad and Mar decided we would go "fork to field" as they say today, and brought in our own cow, chickens and piglets. That was a real homegrown summer! We had some very entertaining pig wrestling in mud going on that you would have had to see to believe.

'It's been a pleasure remembering these things and sharing them."

Chris Dewey, Ranch Manager: "Here is information I found in February 2013 at the Valley County Recorder's office: Received 1960: GRANTOR-Hornback. GRANTEE A. Paul & Caryl L. Cox et al.

'Received 1995: GRANTOR-Judith F. Wiegand, George Wesley Wiegand Jr., George D. Cleminson and Betty C. Cleminson, A. Paul Cox. GRANTEE Caryl L. Cox et al.

'Received 1997: GRANTOR- Caryl L. Cox GRANTEE-Karl E. Giguiere, Victoria L. Giguiere, William T. Casey, Carolyn Knott and Warren L. & Norma K. Nelson Living Trust.

'Received 1998: GRANTOR- Karl E. Giguiere, Victoria L. Giguiere, William T. Casey, Carolyn P. Knott and Warren L. & Norma K. Living Trust. GRANTEE Karl E. Giguiere, Victoria L Giguiere, William T. Casey, Carolyn P.K. & V Giguiere. Received 2005: GRANTOR-William T. Casey et al. GRANTEE-Martin S. Zemitis & Susan B. Smith.

'Received 1967: (C18) GRANTOR-Hornback. GRANTEE Dooley to Scott & Conroy Estate. 'Received 1981: GRANTOR- Scott & Conroy Estate. GRANTEE-Nathan E. Scott et al. Received 1989: GRANTOR- Nathan E. Scott, et al. GRANTEE- Clay and Lois Lacy.

'Received 2008: GRANTOR-Lois Lacy. GRANTEE-J. Michael Paulson, Richard Paulson et al."

Mike Paulson: "Clay Lacy is the reason we ended up knowing about Sulphur Creek. Clay and John Conroy flew over the Idaho wilderness area in the mid-1950s on training missions in their Air National Guard F-86 jets and went back in a small plane to explore and find Sulphur Creek. When Marv moved to Pistol Creek, we followed him."

Robert "Bertie" Boyd: "In summer 2004, I was at the top of the runway when a pilot who was making a routine delivery came up to me in a hurry and handed me a yellow box with formal document stating it contained the remains of Clay Woods, who had died earlier that year in Boise. There was also a note saying that his wife had wished his ashes to be scattered or otherwise placed at rest at the Ranch, where they had been managers. I had not heard of Clay Woods and was doubtful of his association since no one I was able to ask had heard of him either. At that time, everyone at the Ranch was preoccupied with its restoration after the 2000 Fire. However, I also felt it more appropriate to bury his remains in the Ranch cemetery alongside those of others associated with Pistol Creek, so I put the container in our cabin's shed to await my next visit to the Ranch. A few weeks later I returned with my copy of the Church of Ireland Prayer Book, took the box to the cemetery, dug a grave near some rocks in the northwest corner, poured the contains into the grave and read The Order for the Burial of the Dead aloud over it. I then placed at the grave a wooden cross I had made, inscribed "Clay Woods R.I.P."

Steve Kirby: "Bill Brace walked into my office late September 1982. He announced that Dan Plunkett had decided to sell his cabin. I contacted Steve Markstein and Craig Price, who were also excited about the prospect of being part of Pistol Creek. We flew to the Ranch to meet Dan a few days later. It was a rather cold and dreary October, but the Ranch looked fabulous.

'Dan ushered us through the front door. The aroma of the elk stew was simmering in the black kettle pot in the fireplace. Dan wanted $150,000-a fair price, but we were two experienced lawyers and a successful businessman. Dan represented himself.

'We felt bad about the mismatch, but were prepared to take advantage of it. We had a few cocktails and sat down to the great elk supper. We talked about everything, but the purchase price. We slept soundly the three of us dreaming about how fortunate we were and about the hard bargain we would strike. At breakfast, I brought the subject of price up, telling Dan that while we liked the cabin okay, his price was a little high, but we were met by the same pleasant but firm, "Sorry boys." Our plane arrived at 10 a.m. I suggested to Steve and Craig that we bid Dan farewell and walk toward the airplane. I was convinced Dan would change his mind. When we were only a few steps from the plane we turned around and looked back only to see Dan smile and wave with a big "See ya, boys- have a nice trip." We looked at one another, came to our senses and agreed that we better seal the deal before he raised the price. We concluded our purchase with Dan in an amicable way. He could not have been more generous. He held back all of his cast iron cookware, his big game mounts, and a propane deep freezer. Dan asked if he could join us to continue his passion for Elk hunting. We said, "Of course." In the end, Dan left all of his possessions behind, and he never returned to hunt. While the three of us were no match for Dan's bargaining skills, at least we were smart enough not to let the opportunity pass. Before the 2000 fire, along with Braces we hired a bush pilot to fly in and rescue photo albums and all of Dan's game mounts from our cabins. The skins and mounts hung in the Lodge for about two years while we rebuilt the cabin and brought them home to cabin 13."

BIBLIOGRAPHY

1) Alt, David and Hybdman, Donald, *Roadside Geology of Idaho* (Missoula, 1989).
2) Carrey, Johnny and Conley, Cort, *The Middle Fork & The Sheepeaters War* (Backeddy Books, 1977).
3) Carrey, Johnny and Conley, Cort, *The Middle Fork: A Guide* (Backeddy Books, 1992).
4) Cessna Pennant, Fall 1962.
5) Chesterman, C.W., *Field Guide to North American Rocks and Minerals* (Knopf, 1978).
6) Conroy, John Michael, Wikipedia website.
7) Deren, Matthew, *A Forgotten Wilderness* (Donning Company Printers, 2011).
8) Dorward, D.M. and S.R. Swanson, *Along Mountain Trails (and Boggy Meadows)*, (Boggy Meadows, 1993).
9) Dougal, Harold, *Adventures of an Idaho Mountain Pilot* (Self-published, 2012).
10) Dow, G.F., *Every Day Life in the Mass. Bay Colony (in the 1600s)* (Dover Publ., 1988).
11) Fuller, Margaret, *Trails of the Frank Church River of No Return Wilderness* (Signpost Books, 1987).
12) Holm, Richard H. Jr., *Bound for the Backcountry: A History of Idaho's Remote Landingstrips* (Cold Mountain Press, 2012).
13) Montgomery, Payne, Sohn, Thompson, and Toole: *Idaho Wildflowers, Pistol Creek Ranch History* (Greasewood Press, 2018).
14) Pough, F.H., *Peterson First Guide to Rocks and Minerals* (Houghton Mifflin Harcourt, 1998).

INDEX

—A—
Albertson, Joe	68, 69
Arnold, Ray	12, 70

—B—
Barlow, Carl	VII, VIII, 21
Barton, Gene	20, 68
Black, Sear	53
Blake, Kathy	53
Blake, Tad	53
Boyd, Don	66
Boyd, Karen	24, 53
Boyd, Robert "Bertie"	VI, 26, 53, 65, 74
Bolstad MD, Owen C.	36, 39, 40, 43
Brace, Bill	VI, 18, 21, 53
Brace, Maria F. Elosu	53
Brace, Mary Lou	53
Brace, Meredith	VII, 68
Brace, Robert	53
Brace, Robert L. "Rusty"	VI, 12, 17, 19, 53, 68
Brace, William	18 53, 68, 77
Bradley, Durl	20
Brooks MD, Gary	46
Brooks, Marykay	13, 46
Bross MD, Willard, "Bill"	VII, 14, 20, 21
Brown, Barry	53
Brown, Benj. A.	24, 26
Brown, Benj. F.	24, 26
Brown MD, Charles	VII, 14, 16, 20, 24, 26, 53
Brown, Dean	20, 66, 68, 69
Brown, MaryLewis	16, 20, 53
Brown, DeWitt	VI, VIII, 10, 53
Brown, Harold	IX, 36, 46
Brown, Helen	24, 25, 26
Brown, Louie	16, 23, 25
Brown, Patrick	53
Brown, Steve	VI

—C—
Chalupca, Tommy	70
Chaney, Buzz	19, 20, 65
Chapman, John	69
Casey, William	75
Church, Senator Frank	6
Clark, Chuck	53
Clark, Meliss	VI, VII, 16, 53, 65
Cleminson, Betty	75
Cleminson, George	75
Coleman, unknown	20
Collard, unknown	30
Conroy, John Michael	VIII, 14, 18, 22, 23, 24, 26, 68, 75
Coolidge, President Calvin	6
Cox, Caryl	20, 75
Cox, Paul	19, 75
Cox, Caryl	74

Crofts, Arthur	11
Crosby, Jeff	VIII, 17

—D—
DeLong, Betty	20, 68
DeLong, Dick	20, 65, 67, 68
Dewey, Dave	III, 4, 36, 44, 46, 48, 52
Dewey, Chris	VI, 20, 24, 36, 74
Dixon MD, Sherwood	VII
Dixon MD, Trudy Larson	VII
Dooley, Estelle	20, 68
Dooley, Dale	20, 68
Dougal, Harold	VI, VI, 7, 10, 11, 16, 67
Dovel, George	20, 66

—E—
Erkel, Albert	29, 34, 53, 67
Erkel, Donna	29, 53

—F—
Fermoile MD, Kristin Sohn	VI, 12, 38, 44
Fermoile, Mark	VI, X, 15, 39
Flanary MD, Jack	IX, 36
Fleming MD, Charles	VII, 20, 38, 68
Fleming, Chip	66
Fleming, Jan	66, 67
Fleming, Linda	66
Fleming, Shirley	20, 66
Fox, Matthew	53
Fox, Margheria	53

—G—
Geiling, Ross	12, 37
Gerlach, Billie	VI, VII, 20, 53, 64, 66
Gerlach, Clint	VI, VII, III, 19. 20, 53, 64, 67, 66
Gerlach, III, Clint	3, 53
Gerlach, Kim	17
Giguiere, Karl	75
Giguiere, Victoria	75
Giguiere, Victor	74
Gleason, Doug	40
Gleason, Frank	40
Gleason, Hadley	40
Gleason, Jerry	40
Gleason, Langham	40
Grammar, Ben	20, 53, 54
Grammar, Joan	53
Grant, Al	67
Grant, III, Alfred	25, 26
Gubler, Bea	VII

—H—
Hamer, Charlie	11
Haskel, unknown	29
Hawn, Barbara	III, VII, IX, 36
Hawn, Pat	III, VII, IX, 36, 37, 39, 40, 67
Heater, Dewey	14, 65

Heater, Don	14, 21, 65, 66	Patterson, Bill	53
Hogan, Roy	IX, 36, 38, 39	Patterson, Connie	53
Hopkins, Sam	IV, 4	Patterson, Donna	52
Hornback, Barbara	14, 67	Paulson, Allen	20, 66, 73
Hornback, Jackie	69, 70	Paulson, Bob	21 73
Hornback, Marvin	14, 67, 69, 70, 71	Paulson, Mike	VI, VII, 73, 74, 75
Hornback, Patty	70, 72	Paulson, Richard	75

—K—

		Payne, Bill	1, 12, 27, 45, 44, 46, 48, 50
Kelly, Bill	11	Payne, Coleen	46
Kennedy, Robert	70	Pecora, Bill	16, 20, 67
Kent, Cookie	53	Pecora, Dave	VII, 6, 7, 9, 18, 21, 65
Kirby, Hayden	53	Pecora, Kim	18
Kirby, Julie	VI, 13, 16, 53	Pecora, Joan	20
Kirby, Raenel	53	Pecora, Scott	21, 53
Kirby, Steve	VI, 53, 75	Pecora, William	26
Knievel, Evil	27	Pecora, Wilma	53
Knott, Carolyn	75	Perry, Collin	50

—L—

		Perry, Suzi	27
Lacy, Clay	14, 19, 69, 72	Peters, Don	69, 70
Lacy, Lois	20, 75	Peters, Julie	69
Lancaster, Chet	20, 25, 26, 66	Plunkett, Dan	17, 20, 75
Lancaster, John	25, 26, 27, 28, 66, 68	Posing, Floyd	22, 65
Lancaster, Sally	20	Price, Craig	75

—R—

Lukas, Nancy	53		
Lanham, Rex	14	Raenel, Hayden	74
Lanier, H.	20	Richardson (painting)	11, 66
Lind, Ferris	20, 65, 66	Risley, George	IV, 4, 6, 7
		Robbins, Dave	14, 17, 24, 66

—M—

		Russell, Richard	20, 66

—S—

Markstein, Steve	29, 33, 53, 75	Schmitt, Matt	IX, 39, 45, 53
Markstein, Hayden	52	Schmitt, Maximillian	45
Markstein, Raenel	52	Schmitt, Spencer	45
Mattich, Jack	20, 66	Schmitt, Stephanie	53
McCaw, Bruce	53	Schumacher, Jack	67
McCaw, Jolene	53	Schumacher, John	VII, 18, 53, 67
McKeighan, Bruce	IV, VII, 17, 69, 70	Schumacher, Kristine	53
McKeighan, Natalie	69, 70	Schumacher, Scott	VII, 3, 11, 18, 22, 53, 67
McKenna, Bruce	19	Scott, Earl	68
McKenna (McKeighan), Lori	VII, 13, 17, 23, 69	Scott, Dr. Jim	20, 65, 66
McKenna, Patty	21	Scott, Nathan	75
Meyers, Charles	4, 7	Shahin, Ansia	29, 53
Miller, Jessie	36	Shahin, Julia	29, 53
Miller, Paul	36	Slette, Laure	VI, VII, 15
Middleton, Ellen	53	Smilanich, Dennis	VIII, 18, 66
Middleton, Rick	53	Smilanich, Wilma	20
Mitchel, Bill	IV, 4, 7	Smith, Susan	75
Montgomery, Robert	IX, 12, 26, 27, 36, 40, 41, 43	Sohn, Alex	40, 50
"Monty", "Bob"	45, 46, 50, 53	Sohn MD, Anton P.	VII, IX, X, 3, 4, 8, 11, 12,
Moran, Bruce	VI	"APS", "Andy"	13,15, 24,27,29, 37, 39, 40,
			45, 46, 47, 49, 53, 67, 69

—N—

		Sohn, Arlene	VII, 46, 53
Narachi, Elsha	53	Sohn, Bill	12, 39, 40,
Narachi, Michael	53	Sohn, Robert "Bob"	40
Nelson, Norma	75	Sohn, Brady	IX, X,38, 39, 45, 50
Nixon, Jackie	13	Sohn, "Chris"	IX, 40
Nixon, Ray	13, 66	Sohn MD, Eric	IX, 39

—P—

		Sohn, Isabella	44
Patrick, Andy	VI, VII		
Patrick, Scott	24, 25		
Patterson, Alan	53		

Sohn, Kerry	IX, 38, 44, 48, 49, 50	**—V—**	
Sohn, "Liz"	40	Vance, Patty	VI, VII, 20, 68, 70
Sohn, "Mimi"	44	**—W—**	
Sohn, "Peter"	X, 4, 40, 47, 48, 50, 52	Wagner, Dan	25, 26
Sohn, "Phil"	IX, X, 4, 36, 39, 40	Warren, Patrick	52
Sohn, "Rob"	X, 40	Watson, "Wild" Bill	9, 16, 22, 65, 70
Sohn, Sierra	44, 50	Watson, Eleck	IV, 4, 15, 65
Spalding, Quinn	53	Watson, Joella	74
Swanson, Blake	VII, 29, 53	Wayne, Adelaide	IV, 14, 66
Swanson, Kathy	53	Wayne, Bill	IV, 9, 14, 16, 67, 70, 71, 74
Swanson, Sally	53	Weaver, Nancy	53
Swanson, Tad	53	Weaver, Scott	53
Swanson, William	53	Weigand, George Jr.	75
Swenk, Gene	36	Weigand, Judith	75
Swenk, Mike	36	Weigand. Harvey	10, 11
—T—		Weigand, Mrs. Harvey	11
Thompson, Bill	VIII, 9, 12, 27, 40, 45, 46	Whiteman, Gary	68
Thornton, Gail	53	Whiteman, Marv	68
Thornton, Jack	20, 53	Widgren, Bill	III, VII, IX, 27, 29, 36
Thornton, John	53	Widgren, Sonja	III, IX, 36
Tjumslond, Duane	13, 66, 69	Woods, Clay	VI, 19, 26, 64, 74
Tjumslond, Jean	15	**—Y—**	
Toole, L.E. "Gene"	IX, 27, 28, 36, 38, 40, 45, 46	Yupe, Diana	VII
Turner, Daryl	67	**—Z—**	
Turner, H.L.	VIII, 19, 66	Zemitis, Martin	29, 74
Turner, Virginia	20		

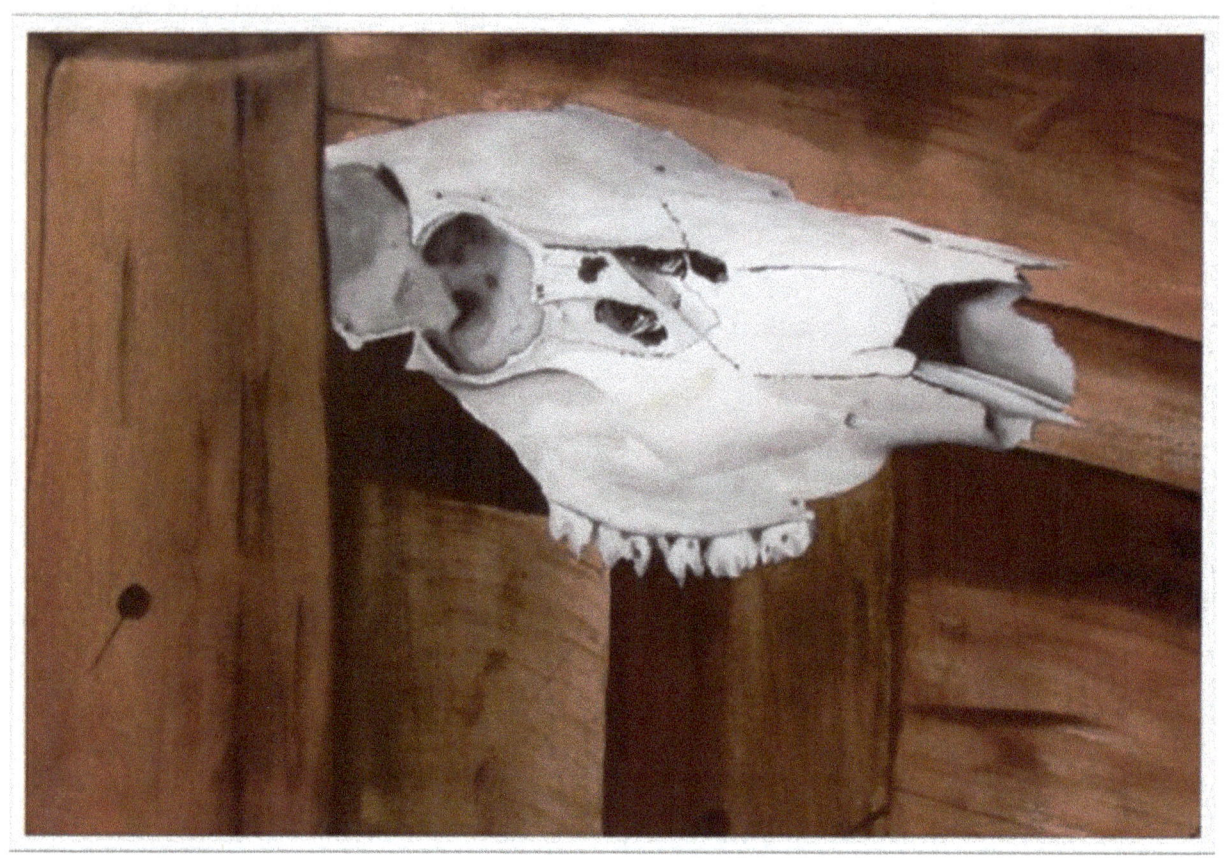

ELK SKULL ON LODGE DECK

Other Titles by Anton P. Sohn MD

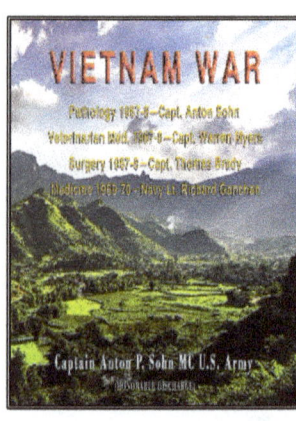

Vietnam War
- Author: Captain Anton P. Sohn MC U.S. Army
- Publisher: TotalRecall Publications
- Hard Cover: 9781648831508
- Number of pages: 110
- Publication Date: 2022

This book records forensic and anatomical pathology at the 9th Medical Laboratory in Saigon from April 1967 to April 1968 of the Vietnam War. I have also included photographs and information from 9th Medical Laboratory Veterinarian Captain Warren D. Myers, 7th Surgical Hospital Surgeon Captain Thomas W. Brady, and MACV-SOG Navy Physician Lieutenant Richard P. Ganchan.

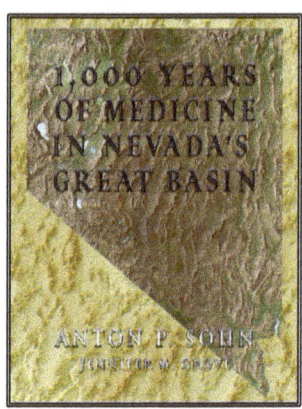

1000 Years of Medicine in the Great Basin
- Author: Anton P. Sohn MD
- Publisher: TotalRecall Publications
- Hard Cover: 9781648831607
- Number of pages: 500
- Publication Date: 2022

This is the story of medicine in the Great Basin and adjacent areas from the beginning of known life in the Great Basin to the twentieth century. The title of this book emphasizes that Shoshone and Paiute Indians lived in the Great Basin for thousands of years. They practiced medicine using all parts of the Greasewood plant and other native plants to treat aliments.

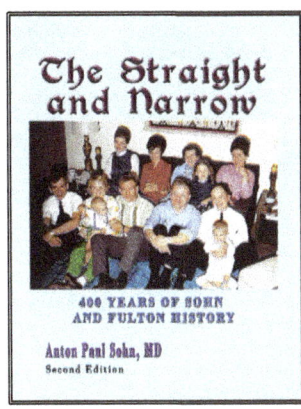

TheStraightAndNarrow

- Author: Anton P. Sohn MD
- Publisher: TotalRecall Publications
- Hard Cover: 9781590951323
- Number of pages: 270
- Publication Date: 2023

    I am writing this manuscript to record family history and tradition, an important part of family pride and awareness. Furthermore, I have enjoyed the opportunity to revisit family and, in some instances, meet relatives for the first time. This document will also create a factual record, although not all information could be verified, and tell a story that is typically American. The story is about families leaving Europe and the British Isles to escape war and famine and to search for and find opportunity in America.

Pistol Creek, Idaho

History and Geology Pistol Creek Ranch Events

- Author: Anton P. Sohn MD
- Publisher: TotalRecall Publications
- Hard Cover: 9781648831881
- Number of pages: 100
- Publication Date: 2023

    The earliest "known life" at Pistol Creek and on the Middle Fork is recorded by pictographs. No scientific evidence to accurately date these images, but archeologists estimate them to be approximately one thousand years old.

    Pistol Creek Ranch History commemorates one hundred and thirty-one years of settlers at Pistol Creek Ranch on the Middle Fork of the Salmon River. The first homesteader came to the area in 1892, but a cabin wasn't built until sometime around 1910. It is difficult to establish the exact date due to lack of records in Valley County where Pistol Creek Ranch is located.

www.ingramcontent.com/pod-product-compliance
Lightning Source LLC
Chambersburg PA
CBHW080628170426
43209CB00007B/1536